100

insects of arkansas and the midsouth

· PORTRAITS & STORIES ·

by Norman and Cheryl Lavers

Table of Contents

For Gareth

Without the patience, encouragement, and hard work of our editor, Erin, and the intelligent approach of our designer, Amy, this book would not have been realized. We thank them sincerely. We want to thank Jeffrey Hoeper for looking at parts of this MS and making useful suggestions. And special thanks to Herschel Raney for inspiring us at the beginning of this century with his website *Random Natural Acts,* and insisting that we get digital cameras.

Introduction

Whether you rarely notice insects, or already spend significant time experiencing the natural world, we have written this book to pique your curiosity as to their astonishing lives. Perhaps you've been revolted by insects in the past, but we hope this book might help you reconsider, and that perhaps you'll pass along the stories that helped change your mind. Just underfoot are some of the great miracles of our world that are almost always overlooked.

The fact is, insects don't need any more enemies; they desperately need friends. And their lives are more intimately connected to our own than we may understand or acknowledge. We hear all about the charismatic insects. Many of us know, for example, that in 2016, the number of Monarchs was down over 70 percent since the mid-1990s. That's terrible enough, of course. But what scientists are also beginning to realize is that the less celebrated species, the ones we don't pay attention to, are disappearing just as rapidly as the Monarchs.[1] The insects that pollinate fruits, vegetables, and wild flowers in spring; the insects that chew up our dead and decaying vegetation and convert it back into soil and new life; the insects that provide us with a host of other services . . . they are quietly vanishing. Insect-eating birds (including most of our songbirds) are also disappearing at frightening rates. Bird populations are declining in sync with the declining rates of the insects which provide their food base. It's the usual causes—clearing the natural landscape to build houses or grow crops, pouring tons of poisons on everything—as the weather becomes ever more unpredictable. Perhaps if we understood the essential roles of these insects better, we'd rethink some of our actions.

We sometimes teach classes on insect ecology for the Arkansas Audubon Society, or do programs for Master Naturalists. Here is how our classes go: After a minimum of formal class work, we all go outside—usually a state park or somewhere similar—to walk and look for insects. The walk is the part we all enjoy most. With so many sharp eyes, we only go a few steps before we spot some interesting insect. We identify it if we can, and usually there is something special to say about the life of that particular creature. A few more steps, and someone will spot another, because insects are everywhere.

You might experience this book like one of our walks, starting with a photograph of an insect, rather than an actual sighting. We'll say something about its life, its strategies, its interactions with other creatures, and then we'll go on to the next image. Themes running through the book include mimicry, warning coloration, the strategies plant-eaters use to avoid being predated, and the strategies predators use to catch enough insects to avoid starving.

The whole finely tuned balance has been negotiated over millions of years: the vegetarians eat enough plants to keep the vegetation from running wild and smothering its own environment, the predators eat enough plant-eaters to keep them from destroying all the plants, the hyper-predators rein in the predators, and the janitors come in and clean up everything left over. In this way, the book is not just an introduction to this insect or that, but an introduction to insects in general, going largely unnoticed, doing all the thankless but necessary work required to keep our world running.

Our hope is that after you read this book you might consider adding some native plants to your garden to attract insects, stop using poisons to eliminate them, and take greater pleasure in the richness of life that is often right before our eyes.

Let's go find our first insect.

---

[1] See Black, Scott Hoffman. "North American Butterflies: Are Once-Common Species in Trouble?" Wings, Journal of the Xerces Society 39 [2016] 314.

Dogbane Beetle (*Chrysochus auratus*)

Coleoptera: Chrysomelidae

Every way this spectacular beetle turns, it flashes greens, golds, and bronzes. It feeds only on dogbane. When you see "bane" in the name of a plant, it signifies that it is poisonous. In fact, dogbane is related to milkweed, the poisonous plant on which Monarch caterpillars feed, making them and the Monarch butterfly they turn into poisonous to anything that tries to eat them. The Dogbane Beetle similarly becomes poisonous eating dogbane. For that reason, it can sit calmly in the open, flashing its colors to the world without worrying about predators. Many brightly colored insects turn out to be unpalatable, so predators immediately find them suspicious.

Lady Beetles (Coccinellidae)

Coleoptera: Coccinellidae

"Ladybird, ladybird fly away home . . ." This one at the top left is just taking off, giving us a picture of beetle anatomy: It has raised the hard protective forewings (the elytra) out of the way, so the soft flying wings can be unfurled.

There are some 500 species of this very familiar insect in America. Many are some color of red or orange with black spots, but many others (black, white, variously patterned) you might not recognize. The one pictured top left is the Multicolored Asian Lady Beetle (*Harmonia axyridis*), which comes in a bewildering number of patterns and colors. It was introduced to our country from Eurasia and is now common everywhere.

Beetles (Order: Coleoptera) are among the advanced insects that have complete metamorphosis. That is to say, when the egg hatches, a larval form emerges which is altogether unlike the adult form, as can be seen in the larval lady beetle on the top left side of the twig, pictured center. When the larva reaches its full growth, it will become a pupa (see be-

low the larva on the left side), an immobile state in which it changes its internal parts so completely that when the adult lady beetle emerges, it is virtually a new animal. (There is an empty pupal case on the right side from which a beetle recently emerged.) Among other changes, the adult has wings and mature genitalia, so that it can move around quickly and find a mate to create the next generation.

The Multicolored Asian Lady Beetle was introduced to this country to help us fight agricultural pests. Indeed, they help us by feeding voraciously on aphids, but we are less pleased with their immense hibernation aggregations, which become a nuisance when they are in our houses or on us (like those swarming Cheryl at a wildlife reserve, top right).

However, here is a lady beetle (bottom left) that is probably too small to ever be a nuisance to anyone, but is worthy of mention because of its wonderful name, The Twice-stabbed Lady Beetle (*Chilocorus orbus*).

Tiger Beetles (*Cicindela* spp.)

Coleoptera: Carabidae

Walking down a path through the woods, you will sometimes see a half-inch long bright metallic green insect running along ahead of you, flying on ahead if your footsteps get too close. This will likely be a Six-spotted Tiger Beetle (*Cicindela sexguttata*), a species which is usually found in openings in woods, often running back and forth on the bare trunk of a large fallen tree. Most of the other approximately twenty species of tiger beetles in Arkansas are found in open country, paths through pastures or open fields, muddy or sandy beaches, or abandoned quarries. The reason they want to be on smooth ground with not much in the way is that, allowing for their size, they are among the fastest running creatures on earth. When a fly or a bee lands on one side of a path, they can run over from the other side, catch the prey in their jaws, and rip it apart before it has a chance to take off.

Whenever you are on a dirt path in the countryside, there will almost always be one or more tiger beetles around. They come in many patterns and often bright colors, but you can always tell them from other kinds of beetles by this mark: with their bulging eyes, their head is wider than their thorax.

Tiger Beetles are predators and their jaws are formidable. Here, at bottom right, is a close up of the jaws of another species, the Hairy-necked Tiger Beetle (*C. hirticollis*).

Their larvae are equally fierce. They live in a burrow in the ground with just the top of their head and their wide-open jaws at the surface, ready to seize anything that comes near and drag it underground. Locally, the larvae are called "chicken-chokers" because it is thought that if a chicken grabbed one and tried to swallow it, it would grip the inside of the chicken's throat and not let go. Children quickly learn that if they put a grass stalk down the hole, the beetle larva will clamp onto it, and they can pull the larva out.

Mottled & Golden Tortoise Beetles (*Deloyala guttata & Charidotella sexpunctata*)

Coleoptera: Chrysomelidae

Tortoise Beetles look like they have had transparent turtle shells placed over their backs. They are common on morning glory plants. They are small and quite beautiful, which is more than can be said for their larvae, which, like many small beetle larvae, protect themselves from being eaten by piling their own excrement over their backs.

Horned Passalus or Bess Beetle (*Odontotaenius disjunctus*)

Coleoptera: Passalidae

You sometimes see these enormous wood-eating beetles lumbering along from one decaying log to another, often with a number of parasitic mites on their head. These mites feed on their bodily fluids but don't seem to harm them. If you accidentally break open an old stump, you might uncover a family group. The adults, male and female, all help care for the larvae, feeding them a slurry of chewed wood covered with cellulose-digesting bacteria from the parents' saliva. This is how the parents introduce the bacteria into the system of the larvae so that the young will be able to digest their own food.

Adults and young have stridulating devices on their bodies and can make up to seventeen separate squeaking sounds, the most elaborate system of sound communication of any arthropod. When they all make their sounds together, it seems to discourage predators.[2]

---

[2] See http://entnemdept.ufl.edu/creatures/misc/beetles/horned_passalus.htm (accessed March 20, 2018).

Eyed Click Beetle (*Alaus oculatus*)
Coleoptera: Elateridae

Click beetles (of which there are nearly a thousand species in North America) have the ability to snap a spine they have into a groove on their underparts. This makes a loud click and propels them several inches in the air, presumably to startle a predator into releasing them, and also to right themselves when they fall on their backs. The huge eye-like marks on this species call attention to it. A similar species we have seen in Costa Rica flies at night, its eye-spots glowing brighter than a firefly's light. They look like they are flying around with their headlights on high-beam. The locals in Costa Rica call them "Fords."

(facing page)

Firefly (*Photinus* sp.)
Coleoptera: Lampyridae

Fire "flies" or, their other popular name, lightning "bugs" are, technically speaking, neither flies nor bugs. They are beetles. If you don't worry too much about keeping your lawn and garden tidy, leave a few scruffy unmowed places and you might have a good population of fireflies in your yard. In the twilight, the males will begin flashing. If there is still enough light to see the individual insects when they are not flashing, it's fun to try to photograph them on the wing just as they flash. Usually, if we get them in focus they aren't flashing, and if we get them flashing, they aren't in focus. The males flash the proper code for their species. If a female, sitting in the bushes watching, is suitably impressed, she will flash back in the proper code for her gender, and they get together. A problem is, females of another genus, *Photuris*, sometimes fake a female *Photinus*'s signal, and when the male rushes over, the *Photuris* kills and eats him.

Whirligig Beetles (*Dineutus* sp.)

Coleoptera: Gyrinidae

On a hot day, in a shaded corner of a lake, you may see a raft of black beetles extending over several feet resting quietly next to the shore. If you come closer, they begin swimming around in slow curves, and if you come even closer, they suddenly start spinning off wildly in every direction. Whirligig Beetles. Nearly everybody recognizes them. They are nocturnal. This is only the place where they sleep during the day. It's part of their defensive strategy to be in a mass and in plain sight, because when they go into their wild gyrations it will totally disorient any potential predator. At night they take out singly and swim around the lake on their own, looking for insects that have accidentally fallen into the water. They swim on the surface, but are also able to dive, taking a bubble of air down with them. A nice feature: They have double eyes—one pair of eyes above the surface so they can see what's above the water, one pair below the surface.

American Carrion Beetle (*Necrophila americana*)

Coleoptera: Silphidae

Don't confuse the similar-sounding name of this common species, the American Carrion Beetle, with that of the rare and endangered American Burying Beetle. But it is in the same lugubrious business: helping to clean up corpses. Its name *Necrophila* means "lover of dead bodies." When they get a whiff of roadkill, they head straight for it. They fly fast and directly, at about eye-height, and something odd occurs. They are suitably black for their job, but note the yellow thorax with a dark patch at its center. On the ground it doesn't look like anything special, but this is the pattern of a bumble bee, a fat black body and hairy yellow thorax with a dark bald spot in the center of it. Until you see this beetle flying, you can't believe how utterly convincing it is as a bumble bee. A bird would hesitate to snatch it out of the air for fear that it had a formidable stinger. This sort of mimicry is very common among insects.

Once it arrives at the kill, it feeds on the abundant fly maggots and lays its own eggs. When they hatch (by this time the other scavengers have picked the body clean), the specialist hatchlings begin feeding on the skin and dried bits of sinew the other scavengers can't handle. Quickly cleaning up what would otherwise be horribly stinking carcasses is one of the innumerable services insects perform for us without our even noticing.

Dung Beetle (*Aphodius* sp.)

Coleoptera: Scarabaeidae

Burying animal droppings underground where they can be converted into rich new soil is another dirty job insects perform for us. There are dung beetles large enough to specialize in elephant droppings, on down to ones that take care of rat droppings and smaller. The dung beetles all look pretty much alike, with a broad spade-like blade along the front of the head, and muscular hind legs that curve out and come together behind them like ice tongs for carrying the dungball. Above is one of the very small ones, carrying a rat dropping to an appropriate place away from the other competing dung beetles, to bury it for its own feeding, or to feed its young, which hatch out of the eggs it has laid in the dung ball. Note, to the dung beetle's left, the two tiny flies (the beetle itself is only 5 mm long) who would like to get their share. Animal droppings are in great demand: They are highly nourishing and already half-digested.

If this unusually colorful dung beetle (above), a Rainbow Scarab (*Phanaeus vindex*), cleaned itself up, it might resemble one of the sacred jewel-like Egyptian scarabs, which are also dung beetles. In Egyptian mythology a god rolls the sun around earth all day, then buries it underground at night where it regains energy and comes up refreshed the next day. The scarab rolls dung around the earth, then buries it underground where it can re-emerge as bountiful crops and a new generation of scarabs.

Fiery Searcher (*Calosoma scrutator*)

Coleoptera: Carabidae

Often when you are following a trail through woods, this large brightly colored beetle will appear, walking at top speed, and quickly disappear ahead of you. They seldom stop even for a moment, which makes them very frustrating to try to photograph. Look at the jaws on this one, reminiscent of a heavy duty pipe cutter. Fiery Searchers are looking for large fat caterpillars. When they find one, they make short work of it with those jaws.

During caterpillar invasion years they become particularly numerous, and you can frequently see these handsome beetles climbing the trunks of trees after caterpillars and disappearing among the leaves.

MILKWEED INSECTS,
OTHER THAN THE MONARCH

Milkweed Leaf Beetle
(*Labidomera clivicollis*)
Coleoptera: Chrysomelidae

Monarch caterpillars feed on milkweed, a poisonous plant. The poison, which does not harm the caterpillars, remains in their systems and poisons whatever eats them. The poison is retained by the adult Monarchs, who advertise this by their orange and black warning colors. A whole suite of other specialized insects also feed on milkweed and become poisonous, and also advertise this fact by being bright orange and black. This beetle, for instance, is the Milkweed Leaf Beetle.

Red Milkweed Beetle
(*Tetraopes tetrophthalmus*)
Coleoptera: Cerambycidae

Here's another beetle that relies on its colors to remind predators that it is poisonous.

Large Milkweed Bug (*Oncopeltus fasciatus*)

Hemiptera: Lygaeidae

In addition to beetles and butterflies, a number of bugs feed on milkweed. "Bug" is a slang term used by most people to refer to insects in general, but when an entomologist uses the term bug, it refers specifically to insects in the order Hemiptera, the "true bugs."

Unlike beetles, the bugs, a more primitive group, only have partial metamorphosis. When they hatch from the egg, they are like a smaller version of the adult. This picture (near right) shows (in addition to an adult at the top of the milkweed seed pod) a number of juvenile Large Milkweed Bugs in almost every size from just-hatched to last stage juveniles. As they grow, they periodically shed their skin to make room for their larger size. With each of these molts (called instars), their wing buds grow a little longer. When they reach their full growth as juveniles, they do not go into a pupal form; they merely shed their skin one more time into a fully adult, sexually mature, fully winged insect.

Wheel Bug
(*Arilus Cristatus*)

Hemiptera: Reduviidae

The "true bugs" of the order Hemiptera, such as this Wheel Bug, are marked by having the basal half of the forewings leathery and the outer half membranous, and instead of biting mouthparts, like a beetle, a segmented beak that folds under the chin when not in use. Also, as we just mentioned for the Large Milkweed Bug, they have partial metamorphosis. Some, like the Large Milkweed Bug, use their beaks to suck plant juices. The Wheel Bug uses its beak to suck the juices of its prey.

If you have a garden (that you don't pour a lot of poisons into), you may have seen this more-than-an-inch-long bug with what looks like half of a buzzsaw-blade on the back of its thorax. The long beak looks formidable and it really is. If you try to pick the Wheel Bug up and it gives you a jab, it will make you yell. They use this beak to kill their prey (chiefly caterpillars) and turn the insides into soup, which they suck dry. A little gross perhaps, but you should admire them as powerful actors in the natural system around us which needs a balance of survivors and population controllers. You might especially cheer for them when you see them slaughtering the big tomato "worms"(actually sphinx moth caterpillars) that would otherwise be eating your tomatoes.

13-Year Periodical Cicadas
(*Magicicada* spp.)

Hemiptera: Cicadidae

Cicadas in general, which are found all over the world, are the large oval-shaped insects that sing and clank and zing all through the long hot days of summer. You see the empty skins (called exuviae) of their nymphs clinging to the bases of trees when you get up in the morning, but the adults that have emerged quickly climb up out of sight, and after that you mostly only hear them singing, hidden in the tops of trees.

However, the 13-Year Periodical Cicadas are easy to see when they emerge simply because there are so many of them. These harmless but rather sinister looking creatures have jet black bodies, blood-red eyes, and bright yellow veins on their transparent wings. Here's their bizarre story: The nymphs hatch from eggs laid in trees, drop down, and burrow underground where they feed by sucking the juice from tree roots. All species of cicadas do this, and most emerge as adults in a year or so. But these periodicals stay underground for thirteen years, and then emerge in scattered places throughout the eastern United States including a number of sites in Arkansas. There are four similar looking species, and in May and June of 2015 all four emerged at the same time in Arkansas. If you were at one of the sites and looked up in the trees, you could

see them constantly flying out from among the leaves, as numerous as bees. Their high-pitched whining hissing sound could be heard, in places, throughout a square mile. They had no fear or avoidance instinct. Birds and other animals could eat them all day until they were sick; there were still millions left.

As with so many insects, their life as nymphs was their real life. The adults had the single purpose of mating and laying eggs, and they did not live long after that. Within days, the dead and dying began littering the ground. After a month

they were all gone, and that particular population at that particular spot will not be seen again until 2028. (Different populations are staggered and may come up in different years.)

A nearly identical set of four species comes out every seventeen years, called, naturally, 17-year Periodical Cicadas, but none of those happen to be found in Arkansas. This remarkable phenomenon occurs (periodically) throughout eastern North America, and nowhere else on the planet.

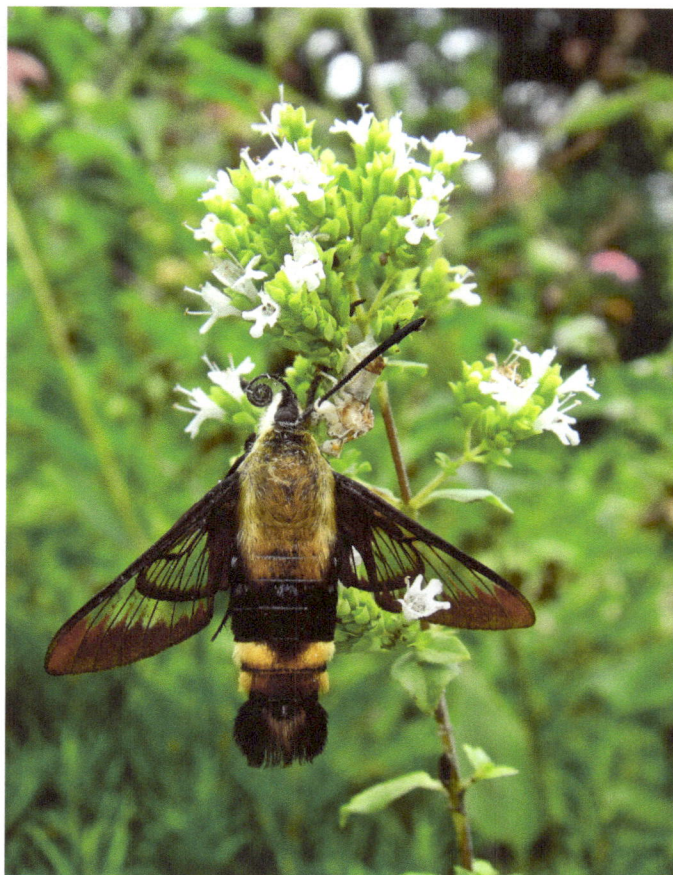

Ambush Bug (*Phymata* spp.)

Hemiptera: Reduviidae

Sometimes you see a flowering bush with a butterfly nectaring at practically every flower. If you look carefully, there will often be one flower from which a dead butterfly is hanging. Go close, look even more carefully, and you will see that something like a miniature gargoyle is holding the butterfly by the tongue or by the antenna in muscular forelegs while it sucks out the butterfly's juices.

This will be the tiny Ambush Bug. In the pictures above, it is shown from different angles, in two colors. But it is much more likely to be concealed invisibly in the heart of a flower, waiting for virtually any of the nectaring insects to land, no matter how big. Its beak injects a powerful fast-acting venom, so there is not much struggling. Don't try to pick this bug up with your fingers, or it might give you a good poke!

If you are not yet properly impressed with the Ambush Bug's big-game hunting weaponry and strength to hold heavy weights, here it is above, this tiny almost invisible thing (maybe a quarter of an inch long), with a relatively enormous Snowberry Clearwing moth. The moth was pretending to be a fierce bumble bee, but the mimicry didn't help it against this foe. (Note, however, if it is not a gargoyle, but a spider with its long arms around the prey, then it will be a Crab Spider, an equally formidable hunter with a similar ambush *modus operandi*.)

Bee Assassin (*Apiomerus crassipes*)

Hemiptera: Reduviidae

Pictured left is another of the predatory true bugs. The Bee Assassin hangs out around flowers in order to waylay bees, on which it feeds. In this image, it has caught a small solitary bee. Note that its front legs are much stouter than its other legs and are covered with short stiff hairs that help it to hold a violently struggling insect that will be trying to maneuver its stinger around to defend itself. It's surprising how many predators specialize in dangerous biting or stinging prey. Perhaps such prey relies on its weapons, and is otherwise not as wary as it should be, and so easier to catch.

Leaf-footed Bugs (Coreidae)

Hemiptera: Coreidae

There are about eighty species of Leaf-footed Bugs. These rather large insects feed on various plants and usually can be recognized by the leaf-like flanges on their hind legs. The Eastern Leaf-footed Bug (*Leptoglossus phyllopus*), left, recognizable by the white band across its wings, is a common and typical example.

The burly *Acanthocephala femorata* is less typical. With those huge hind legs they battle with other males to keep possession of territories attractive to females.

23

Giant Water Bug (*Lethocerus* sp.)

Hemiptera: Belostomatidae

Here is the darkest nightmare for a small fish, frog, or backswimmer. This bug, which hides under water in the mud or vegetation, is over two inches long. The muscular forelegs, coming out of nowhere, can grab prey in a second. The beak, out of sight underneath the Water Bug in this picture, is full of venom to inject into prey, with a hollow point to suck its juices. In addition to being a swimmer, the Giant Water Bug is also a powerful flier. At night they evidently go prospecting for other ponds, for many are attracted to streetlights and end up the next morning stranded in city gutters, which is where we discovered this one.

Rock-loving Grasshopper (*Trimerotropis saxatilis*)

Orthoptera: Acrididae

The order Orthoptera includes grasshoppers, katydids, and crickets. A number of katydids and crickets are carnivorous, but grasshoppers are the grazing animals of the insect world, the equivalent of the antelope on the African plain. The things, in other words, that everything else eats. Therefore evolution has pushed them not towards better weapons, but towards better evasion tactics. Powerful long leaping legs for one, camouflage for another. This grasshopper is one of the supreme examples of the latter. It flourishes on lichen-covered limestone glades such as are commonly found in the Ozarks, its pattern imitating the shapes and colors of the lichen to perfection. People who have noticed this grasshopper spontaneously call it "the lichen grasshopper." The Seven Hollows Trail at Petit Jean State Park is a good place to see them.

Differential Grasshopper (*Melanoplus differentialis*)

Orthoptera: Acrididae

This handsome, very common grasshopper (featured above) is easily recognized by its waxy body and the black herringbone pattern on its hind legs. With its large size, big appetite, and occasional abundance, it can become a pest in gardens. Unlike the true bugs, grasshoppers have chewing mouthparts rather than a sucking beak, but like the true bugs they are also an insect with partial metamorphosis. (Partial or incomplete metamorphosis includes only three stages: egg, nymph, and adult.) The adult here is in the background, and in the foreground a partially-grown nymph, its pattern not fully out yet, and only small wingbuds developed.

Carolina Grasshopper (*Dissosteira longipennis*)

Orthoptera: Acrididae

If you go walking dirt paths in open country in the summer, one of the most common grasshoppers you will see is the Carolina Grasshopper. It will be sitting out on the trail in plain sight, a large grasshopper with long wings, more or less the color of whatever soil it sits on. Look with your binoculars and you will see a pale area just before the tips of the wings where light is shining through. When you get close enough to make it fly, you will note that it has mainly black wings, which confirm your identification.

You will often see the male hovering in one place about four feet high, displaying his flying prowess for any nearby females. You can easily be fooled that it is a black butterfly you are seeing. When these large flying grasshoppers are part of insect collections, they are pinned with one wing opened so you can see the butterfly-like marking. Once, after we made a long highway drive, we found that one had collided with our car and gone up into the grill where it died in its flying position, so we kept it as a perfect specimen to show something of the anatomy

of a grasshopper, and it is now featured at the lower left. The wings that we see when a grasshopper is sitting are the leathery forewings that protect the body like a beetle's elytra protect a beetle's body, and like a beetle's elytra, they are no good for flight. In flight the grasshopper, like a beetle, raises these forewings to get them out of the way, then spreads the soft hind wings for flight. The translucent yellow border to the Carolina Grasshopper's wings allows the light to shine through, causing the pale-appearing band near the end of the closed outer wings.

Entomophaga grylli, a fungus that attacks grasshoppers, sends its filaments all through the grasshopper's body, eating it from the inside out. Just before this Carolina Grasshopper's death (top left), the fungus took over its brain (turned it into a "zombie" is actually the scientific term), and made it fly to the top of a plant stalk, where the fungus spores have the best chance to circulate and find another grasshopper victim.

American Bird Grasshopper (*Schistocerca americana*)

Orthoptera: Acrididae

The bird grasshoppers (Schistocerca spp.) are among our largest and most beautiful grasshoppers. With their size, their long wings, and their powerful flight, they are reminiscent of birds. This one is a common species but you will seldom see it, as it will be down in tall grass, feeding and staying out of sight. But if you walk off the trail you are likely to flush one, which will startle you with its size. It will fly directly away from you for twenty-odd feet, and then instead of landing back on the ground like most grasshoppers, if you are in a lightly wooded area, at the end of its flight it will curve up and land in a tree about ten feet above the ground. We have seen this behavior so often and so regularly that we believe it is enough by itself to identify the species. If you watch where it lands in the tree, it will probably be in plain sight, and if you have binoculars you will be able to examine its attractive markings.

Handsome Trig (*Phyllopalpus pulchellus*)

Orthoptera: Gryllidae

This is a kind of cricket, but it goes to great effort to look like something fiercer, using every dodge it can think of to keep from being eaten. Katydids, for instance, can look exactly like leaves, grasshoppers have exactly the colors of the dirt they sit on, both can leap distances, and fly. The Handsome Trig makes itself obvious to see, but more dangerous looking. The female in this picture, to begin with, has her wings folded in such a way that instead of soft gauze, they look like the hard elytra of a beetle. The long black maxillary palps are oddly flattened to perhaps look like a beetle's biting jaws. In addition, of course, the overall color scheme—black, white, and red—are the aposematic (warning) colors of something bad-tasting, or with a sting or venomous bite.

Oblong-winged Katydid (*Amblycorypha oblongifolia*)

Orthoptera: Tettigoniidae

This is one of a group of rather similar looking katydids. In most of these species from time to time, instead of their camouflage-green coloring they come out in a bright pink or sometimes a lemon yellow form. I don't think anyone knows why, but apparently enough of them survive to breed that the gene for these colors continues to exist. They are sometimes thought to be a great rarity, but if you are watching for them, in most years you can see one or two.

True Katydid (*Pterophylla camellifolia*)

Orthoptera: Tettigoniidae

Above is the true katydid indeed, the one that calls "katydid, katydidn't" all night long, the raspy sound coming through screened bedroom windows on hot summer nights. There are lots of other night-creature sounds, but this one dominates. There are a number of katydid species that have leaf-shaped forewings, but all the others have hindwings that are longer than the forewings, so their pointed tips extend a little beyond the rounded tips of the forewings. On the True Katydid alone they don't peek out, so the final wing shape is rounded. For that reason you can always recognize it. The other katydids, on the other hand, are very hard to tell one from the other.

Straight-lanced Meadow Katydid (*Conocephalus strictus*)

Orthoptera: Tettigoniidae

Meadow katydids are small often rather slender katydids that can be numerous in fields of calf-height grass. Several species of meadow katydid can occur together and they can be very difficult to separate. In many cases the male genitalia need to be examined, and the females are nearly impossible to distinguish. But in the case of this species, the female (left) is easily identified by the great length of its sword-like ovipositor.

Northern Mole Cricket (*Neocurtilla hexadactyla*)

Orthoptera: Gryllotalpidae

It is astonishing how much these strange insects above resemble real moles, their "moleskin" skin, their heavy digging claws, and all of their behavior. They live in the moist soil at the edges of ponds or seeps, and spend almost all their life just underground (shallow enough that they sometimes leave a miniature mole-like ridge where they have been plowing along). Their digging is in search of worms, on which they feed. Just like real moles, you some-times catch them walking around above ground, but these "moles" are only a little over an inch long.

Common Green Darner (*Anax junius*)

Odonata: Aeschnidae

The Common Green Darner is the most commonly seen, very large dragonfly. If you get a good look in good light you will notice the thorax is green. Often, they rest in trailside tall grass a foot or so off the ground and you can sometimes accidentally surprise them into flight. The one you chase up might soon land again, and this time, since you can see it in advance, you might be able to sneak up on it so closely that you can take a nice picture of it. The mature male has the blue abdomen, the female a purplish abdomen. Both always have the green thorax, and from close up a bull's eye pattern on the forehead. This species is migratory, so that you will see noticeable increases of them in spring and fall. Those in this picture are in the "mating wheel" that dragonflies and damselflies use. The male uses claspers at the tip of his tail to hold the female by the back of the neck. She has her genitalia at the tip of her abdomen, and she brings them up to the base of the male's abdomen, where he has his.

Swamp Darner (*Epiaeschna heros*)

Odonata: Aeschnidae

Naiads, the larval forms of dragonflies, creep around slowly underwater in dense aquatic vegetation or tangles of mud and roots. They are usually exactly the color of their surroundings, and when a small living creature comes near, their jaws (on a sort of hinge) shoot out some distance and grab their prey. After perhaps a couple of years for a big creature like the Swamp Darner, the naiads crawl out of the water onto some emergent plant or stick, the middle of their back cracks open, and the adult dragonfly pulls itself out.

Common Whitetail (*Plathemis lydia*)

Odonata: Libellulidae

Above is a handsome dragonfly, the fully mature adult male unmistakable with its solid bluish-white abdomen. The female has the same basic marking, but without the bluish highlights, which makes it a much duller insect. When the male first emerges from the water and takes on his adult form, he is marked like the female (see image at right), and there is a reason for this. For the first few days his wings and outer integument are still soft and vulnerable to injury. His genitalia are not yet functional. So he can hide out from the other males disguised as a female for a few days until he is ready. The males are very aggressive and territorial. They stake out their bit of water and fiercely drive off any other males of their species that come near, at the same time trying to corral any female that enters their territory. It's rough and tumble.

But slowly, as the new male matures, his body becomes pruinescent (a waxy bloom exudes from the cuticle), which is to say it develops the bluish marking on the abdomen that identifies the dragonfly as a male. Not all, but many other species of dragonfly pruinesce according to their own specific pattern. Pruinesence is the sign of his readiness to enter the battle to fight off males and attract females.

Common Pondhawk (*Erythemis simplicicollis*)

Odonata: Libellulidae

If you are a fisherman, or otherwise spend time in wetlands, at a certain time of year you will notice that the ground is almost covered by dozens, possibly hundreds, of medium-sized green dragonflies. Go out a few weeks later and they will seem much less less obvious. What has occurred is that all the males will have turned blue (another example of pruinescence). There are lots of kinds of blue dragonflies, but you can still recognize this species because the blue males will have kept their green faces. These are very formidable hunters, catching quite large prey, including other dragonflies, even including other Common Pondhawks. They are especially fond of butterflies. The green one here is eating a Silvery Checkerspot. They sometimes seem to follow us, like cattle egrets follow cows, catching all the insects we chase up.

Wandering Glider (*Pantala flavescens*)

Odonata: Libellulidae

In the background of this picture above, a Black Swallowtail caterpillar is just beginning to form its chrysalis. You can be excused if that caught your eye before you looked at this pale, rather nondescript dragonfly. But the dragonfly is a marvel. In high summer on a still, humid day, if you look out over rice paddies or open meadows, the sky may be filled with dragonflies from top to bottom. These are swarms of thousands, and if you can get the right angle on the light, you will see that they are all yellowy-brown in color, and in fact they are all this species (sometimes mixed in with a close relative, *Pantala hymenaea*). The Wandering Glider is the most numerous dragonfly you will ever see. Add to that the fact that this species can be seen all over the world, that ships in mid-ocean report large flights of this dragonfly, and finally, it has recently been discovered that it makes a long annual migration virtually across the Indian Ocean, a migration much longer than the Monarch's.[3]

---

[3] See http://www.ted.com/talks/charles_anderson_discovers_dragonflies_that_cross_oceans (accessed March 20, 2018).

Fragile Forktail (*Ischnura posita*)
Odonata: Coenagrionidae

Damselflies are a suborder of the dragonflies, and are the more primitive of these very primitive insects. They are smaller, slighter, and stay down in the vegetation more than dragonflies. Almost all of them fold their wings over their backs rather than holding them at right angles to their bodies like dragonflies. They are no less beautiful than dragonflies, many with very bright colors, especially blues and reds. They are, on their smaller scale, equally fierce hunters, even, when the opportunity arises, catching and feeding on other damselflies. It is not uncommon to see hundreds in a small area along a lakefront, all in mating wheels. If you look carefully, you may see many of the females laying their eggs. They use their ovipositors to make holes in underwater aquatic plants in which to insert their eggs. The male generally holds onto the female while she backs down into the water, sometimes entirely out of sight, to do the laying.

The damselfly pictured above is the smallest member of the dragonfly order (Odonata) in North America, less than an inch long. It is present in gardens even far from water. It is so small and slender it can disappear from view merely by altering its angle. But if you get a good look, the male and female are both instantly recognizable by the pair of exclamation points on their thoraxes.

Ebony Jewelwing (*Calopteryx maculata*)

Odonata: Calopterygidae

Find a small creek with slowly moving water at the base of steep banks dense with vegetation. If a walking bridge crosses the creek, stop there and look over the side. Before long you will see these very large damselflies, sitting on leaves near the water, or slowly flying around with their big velvety wings, displaying to the females. The males have the metallic green bodies. The females are duller, and have white stigmas on their black wings.

Scorpionfly (*Panorpa nuptialis*)

Mecoptera: Panorpidae

Scorpionflies look like monsters from Star Wars. They get the scorpion part of their name from the wicked looking tail on the male, but that is only a harmless sort of grip to hold the female in place while they are mating. This species takes the place of vultures in the insect world. They feed on dead insects, and help to keep that surprisingly messy part of the world cleaned up. The females will not mate, however, without a gift of fresh meat. Flattened, dried out roadkill grasshoppers—or whatever their daily boring fare is—won't do. So sometimes the males get a little desperate. We have heard of the males going into spider webs and stealing fresh-caught flies from the jaws of the spider, and could scarcely believe it. But then we saw the one in this photograph. We were getting our cameras ready to take a close-up of a largish jumping spider with a fly it had just caught when this big dragon-like beast swept in, bluffed the spider back, and stole the fly.

Hangingfly (*Bittacus* sp.)

Mecoptera: Bittacidae

Above is a relative of the scorpionflies. It is much less formi-dable looking, a more spindly creature looking rather like a crane fly, but has its own novel way to capture its prey. The male (pictured here) gets down low in the foliage and hangs by its front legs. But look at its hind legs dangling below it, with their complicated spiny feet. They are like loaded mousetraps, and if a moth or other soft creature in flight brushes against these feet, they snap around it in a flash.

The hanging fly is not hunting on his own behalf. It turns out he can't attract a female to mate with either unless he has a meaty gift to present to her, for her to eat while they are mating. Many insects catch their prey with grasping forelegs, but the hangingfly is one of the relatively few insects that uses predatory hind legs.

Carolina Mantis
(*Stagmomantis carolina*)

Mantodea: Mantidae

The praying mantis of course is the insect most famous for its predatory forelegs. Once those muscular arms have enclosed some hapless insect in their Iron-Maiden grip, there is not much it can do. The mantis uses its delicate mouth to eat off the head, which quickly ends any struggles.

The females have a reputation for eating the males right after mating. More often than not this is undeserved. At any rate, we personally have seen lots of matings and they all ended peacefully, and the males don't seem to have heard the rumor. Above three males at once are struggling for the privilege to mate with this female.

At the end of the year mantises lay their characteristic egg nests, which become readily visible when the leaves come off the shrubbery. If all goes well, in the spring little louvers will open at the bottom of the nest and baby mantises, miniature versions of the adults, will come spilling out. But the eggs are vulnerable, no matter how fierce the mother. Above left, a parasitic braconid wasp has found this nest and is laying its eggs within it. The wasp larvae will hatch quickly and begin eating the eggs inside.

Giant Walkingstick (*Megaphasma denticrus*)

Phasmatodea: Heteronemiidae

Walkingsticks are such good mimics of "sticks" that they are usually difficult to spot, but they occasionally have a population explosion which makes them much easier to find. At such times, you may hear a pattering sound as their eggs are dropped singly into the leaf litter from overhead trees. The eggs look very much like seeds. There are smaller species around that you may see more often, but this one is noteworthy by being the longest insect in North America, sometimes reaching seven inches. If you keep them for pets, sometimes they will wake you up at night with loud clashing sounds as the males battle each other, slamming their forelegs together almost like elk with their antlers.

Northern Walkingstick (*Diapheromera femorata*)

Phasmatodea: Heteronemiidae

To the left is a more common walkingstick, which you are more likely to see.

Green Lacewing (Chrysopidae)

Neuroptera: Chrysopidae

These tiny and frail creatures are in the order Neuroptera, named for the way the wing veins are laid out like a pattern of nerves. The green lacewings have a distinctive way of laying their eggs: They set them up on little stalks, so that ants walk along the twigs without realizing there are delicious little bags of protein just over their heads.

Their fierce larvae are called "aphid lions." If you have an outbreak of aphids in your garden, before you reach for the spray, sit down next to the aphids and watch them with close-focusing binoculars, or take pictures of them with your macro close-up camera and try to see what's going on. There is a whole industry of creatures which feed entirely on aphids usually dealing with your outbreak pretty quickly, whereas if you poison them, the aphid predators will be gone, and the aphids will soon come surging back. See, towards the end of the book, the section on aphids.

Above left, for instance is a green lacewing larva spearing an aphid. Its scimitar jaws are like hollow needles, so it can suck the aphid dry, then go on to the next aphid, and so forth all day.

Mantisfly (Mantispidae)

Neuroptera: Mantispidae

Let's put this related insect right here, because in size (ca 15 mm) and shape, the mantisfly somewhat resembles a lacewing, except, very improbably, it has (on a miniature scale) a praying mantis's mighty grasping arms in front. We sometimes keep these as pets, and they love catching and eating mosquitoes.

There are several species. They all have the Lucy-in-the-Sky-with-Diamonds eyes, and many of them mimic wasps in their markings. Their life history is interesting. When their eggs hatch, the mobile larvae go off looking for big spiders. If they find one, they get on it and stay with it until it lays its eggs (or, if it is a male, until it mates, at which time the larva jumps off the male onto the female). The spider spins a bag to put its eggs into, except the mantisfly larva jumps in the bag first. As soon as the spider lays its eggs in the bag and weaves the nest tightly shut, the mantisfly larva begins consuming the spider eggs. When it has eaten them all, it makes its own cocoon inside the tough protective spider egg-nest, and in the spring instead of 100 spiderlings, a new mantisfly emerges.

Eastern Dobsonfly
(*Corydalus cornutus*)
Neuroptera: Corydalidae

Here's a lovely big insect, another member of the Order Neuroptera (nerve-winged), referring to the wing veins which appear to be laid out like nerves. It's another of those insects that lives the major part of its life as a larva (the famous "hellgrammite"), this one living under water, where it hunts tadpoles, fish, and other aquatic prey. The amazing adults are sort of a waste as they don't eat but only last for a few days, until they can mate and lay eggs. The picture with the hand for comparison shows how large the adults are—not all insects in the order Neuroptera are tiny frail things.

To the far left is a closer picture of the immense jaws of the male, which evidently are used to hold the female while mating and are not very effective as weapons. The next is a picture of the female, and you can see her jaws are more business-like, and can give a very hard nip in self defense.

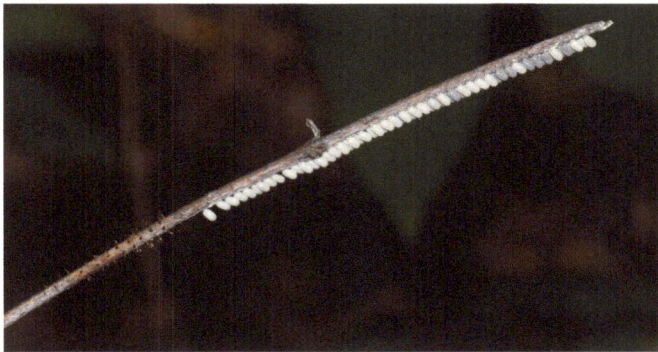

Owlfly (Ascalaphidae)

Neuroptera: Ascalaphidae

These curious creatures also belong to the neuroptera. They resemble dragonflies that have immensely long antennae. We always see them during the day in this pose above, but evidently at night they become fast-flying predators, very like dragonflies.

At certain times of the year, once you learn how to recognize them, you see their characteristic eggs quite commonly. They are laid in two rows at the end of a plant stalk. Ants, which might eat the eggs, can't approach them because little false eggs have been laid in the way which contain some-

thing disagreeable to the ants. It is fun to break the stalk off, take it home, and set it in a glass on the kitchen windowsill where you will see it every day. The eggs will begin hatching, and things with mandibles like mastodon tusks will begin appearing. At a certain point dozens will be out, hanging from the stick. Put the jar out in your garden over leaf litter, and after several days they will begin dropping off, one by one. They wander off in the litter to a good place to sit with their jaws cocked open to wait for whatever hapless small arthropod comes by.

False Crocus Geometer
(*Xanthotype urticaria*)

Lepidoptera: Geometridae

We don't know what these attractive moths have done to deserve their notorious name. And, for that matter, even people who claim they have seen a "true" Crocus Geometer don't always seem very sure of themselves. But whatever the false one's personal morality, it is an often seen, pretty, day-flying, almost-always-mating moth. Once you spot a pair lying in plain sight in the grass at your feet, you will not forget it, nor will you forget their unfair name, associated with the basest treachery.

Cheryl has just done some research and discovered that the DNA of the Crocus Geometer and the False Crocus Geometer is identical, which is to say, by some rules anyway, they may after all be the same thing.

Eight-spotted Forester
(*Alypia octomaculata*)

Lepidoptera: Noctuidae

The Eight-spotted Forester is a very handsome and commonly seen day-flying moth. Most moths are secretive, but this one wants to be seen. Black, white, red, and metallic blue is a color-scheme well known throughout nature—to birds and lizards, perhaps insects and spiders, even human beings (think of bees and wasps: we see black and orange or black and yellow bands, aposematic coloration, and we know not to touch). Some combination of these colors means "Don't touch me, I have a nasty bite, or an envenomed sting, or I taste terrible to the point of being poisonous." Sometimes insects are just pretending with their colors, and can't hurt you or don't taste bad, but who wants to take the chance?

White-marked Tussock Moth (*Orgyia leucostigma*)

Lepidoptera: Erebidae

If you are not accustomed to looking at insects, this might be a very weird one to begin with. It's the caterpillar of a drab and unremarkable moth, but moth caterpillars are often more interesting looking than the moths they become, and that is certainly the case here. The caterpillar above is quite neat with its colored hairs going in different directions. Caterpillars are furry for a number of reasons. For one, something very hairy is difficult and unpleasant for a bird to swallow. It is also trickier for a parasite to attach its eggs to. But finally (and remember this whenever you are tempted to pick up a furry caterpillar), the hairs are often stinging hairs, or hairs covered with chemicals you may have an allergic reaction to. The latter is true of this one. When it sees trouble coming, it curves its back up and rubs its body hairs over the two red glands near the end of its body, coating them with a material very unpleasant to sensitive skin. This protects them well enough that a number of caterpillars that lack the defensive glands nonetheless mimic all the hairs so well that they are also called tussock moths, but the present species is called the "true" tussock moth.

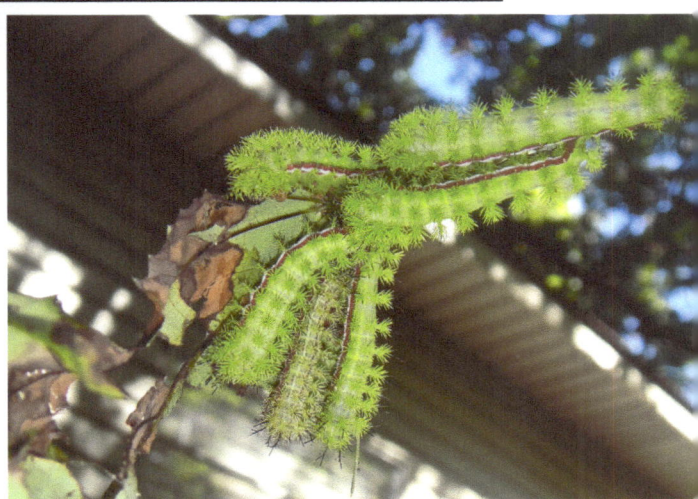

Io Moth (*Automeris io*)

Lepidoptera: Saturniidae

Contrary to what we just said, these caterpillars turn into very attractive moths, members of the American Silk Moth family. But what we said about their bristly hairs is true: If you accidentally brush against these, you will feel like you have just walked through stinging nettles, and you will even get a slight rash.

Geometrid Caterpillar (Geometridae)

Lepidoptera: Geometridae

We have a small Bald Cypress in our front yard. We had trimmed a slim branch off the base of the trunk the previous year, so it was quite reasonable that the tree would send up a new shoot from the same place the next. It was there all day, but the next day it was in a slightly different position, so we looked more closely, and realized it was a geometrid—a "measuring worm, inchworm, or looper"— one of the group of moth caterpillars (very difficult to identify as to species) that can hide in plain sight because of their ability to look exactly like a bare twig, and to hold that position hour after hour. But this one did it more perfectly than any other we had ever seen.

However, we played a trick on it. We knew it had to feed some time, and that time must be at night. We came out after dark, and took a picture of it by flash. Sure enough, it was chomping away on the fresh spring cypress foliage

Robin's Carpenterworm (*Prionoxystus robiniae*)

Lepidoptera: Cossidae

This is a big heavy moth. If you are standing at the porch light looking to see which moths have been attracted, and if this one is flying around, it will almost hurt if it crashes into you. The moth is named for its caterpillar, which, equally big and powerful, bores holes into living trees large enough to damage them. It is related to the "worm" found at the bottom of tequila bottles.

Arkansas Clearwing (*Synanthedon arkansasensis*)

Lepidoptera: Sesiidae

Moths are emblematic of softness, helplessness. Not the sort of reputation that discourages predators from going after you. So a number of moths pretend to be wasps or bees, creatures with venomous stings, to protect themselves. This one, for instance, has black and yellow (aposematic) bands to suggest a yellowjacket, long heavy antennae to resemble the long antennae of most wasps, and most radical of all, clear wings, getting rid of the colored scaly wings that epitomize a moth. For this reason, the many families of wasp-mimic moths, though often unrelated to each other, are called "clearwings." The second part of this one's scientific name, arkansasensis, meaning "from Arkansas," tells you the state in which it was first discovered and described.

Snowberry Clearwing (*Hemaris diffinis*)

Lepidoptera: Sphingidae

Here is a day-flying Sphinx moth that hovers, hummingbird-like, before flowers to drink their nectar. This one is also a clearwing, though unrelated to the Arkansas Clearwing shown earlier. The clear wings here are part of its disguise as a bumble bee, with its black and yellow coloring. This disguise evidently gives it some protection from predators, because these are abundant in our garden. But there are some costs to it as well. Other moths hit the very sticky black-and-yellow garden spider webs and the loose scales on their wings come off and are left behind as the moths wriggle free. The Snowberry Clearwing, with no wing scales, is caught.

Polyphemus Moth (*Antheraea polyphemus*)

Lepidoptera: Saturniidae

These spectacular big moths come to our lighted windows at night. This is a pair mating, the male on the left showing his hugely plumey antennae which can detect the tiniest mile-away trace of the female's pheromone, the female's much reduced antennae showing on the right. She doesn't have to find him; he has to find her. The caterpillars do all the feeding. The adults do all the mating and egg-laying, but they must be quick, because they have no feeding mouthparts, and only live for a few days.

White-lined Sphinx Moth (*Hyles lineata*)
Lepidoptera: Sphingidae

This is one of the commonest day-flying sphinx moths, a big confident animal hovering in front of a flower, its wings a blur (but here stopped by the flash). In a year when they are abundant, we get an occasional phone call from someone who thinks they have seen a new species of smaller hummingbird. Whenever you see a picture of a sphinx moth hovering before a flower, it is almost always this species. First, they are quite beautiful. Second (we shouldn't be giving away these photographer's secrets), they pose tamely in front of the flower, holding their pose for several seconds, giving you all the time in the world to focus and shoot. And, of course, virtually all the other sphinx moths come out in the middle of the night. Look at the long tongue, getting ready to be placed with perfect accuracy down into the tube of that flower.

When they want to rest they can settle down in a tuft of grass and disappear in one second.

Pipevine Swallowtail
(*Battus philenor*)

Lepidoptera: Papilionidae

The caterpillar of the Pipevine Swallowtail feeds on pipevine (Aristolochia), which is very poisonous. The caterpillar manages to sequester the poison in its tissues, where it does not damage the caterpillar but makes it poisonous to anything that tries to eat it. But of course that doesn't do it any good unless potential predators know it is poisonous before they take a bite, so it advertises its bad taste by being dressed in aposematic ("warning") colors—black and red—and by looking rather like a centipede, the appendages on the sides like multi-legs. Whenever you see a pipevine plant winding its way up some tree, it's worth searching through its leaves to see if you can find these huge black caterpillars.

The butterfly that emerges from these caterpillars is still poisonous, and advertises its distastefulness in its own suite of aposematic colors—black, white, red (on the underside), and metallic blue—which are studiously copied by nearly half a dozen probably nonpoisonous other species of butterflies. Namely: Spicebush Swallowtail (Papilio troilus), Black Swallowtail (Papilio polyxenes), Eastern Tiger Swallowtail in the black form of the female (Papilio glaucus), Diana Fritillary female (Speyeria diana), and Red-spotted Purple (Limenitis arthemis).

Eastern Tiger Swallowtail (*Papilio glaucus*)
Lepidoptera: Papilionidae

Here is the pattern of the Eastern Tiger Swallowtail male, and probably fifty percent of the females.

The other fifty percent of the females have this dark form, which mimics the poisonous Pipevine Swallowtail, in the hopes that predators will avoid it, thinking it is poisonous. The question arises: Why don't all the females adopt this protective black form? It turns out that the male Eastern Tiger Swallowtails are more likely to mate with recognizable black-and-yellow females. So the choice is: more sex or more protection.

The Eastern Tiger Swallowtail caterpillar has its own dodges. Its fake eyes give it a convincing snake head, and it can stick out a rather convincing snake "tongue," which in fact sprays a poison gas. The "tongue" (technically called an osmeteria) is a chemical defense that all swallowtail caterpillars possess.

Long-tailed Skipper (*Urbanus proteus*)
Lepidoptera: Hesperiidae

The skippers are a subset of the butterflies. They are generally smaller with stockier bodies and short, pointed wings which enable them to fly so fast that birds won't even try to chase them. This particular species is a rarity in Arkansas from farther south, but occasionally has an invasion year and can appear anywhere in the State. It's such a glorious butterfly, you don't want to miss it. It's very easy to identify with its long tails and blue-green back.

Underwing Moths (*Catocala* spp.)
Lepidoptera: Erebidae

When you begin to get into the real heat of summer, you can enter a wood and it will seem to be devoid of life. But from time to time as you walk along, you will flush a large moth from near the base of a tree, and it will fly off fifteen or twenty feet before curving around behind another tree and landing low down on the far side of the trunk from you. If you walk up to that tree very quietly, and peek around the trunk, either the moth will fly off and you will have to follow it to another tree, or it will stay still and after a search you will finally spot it, blending perfectly into the bark of the tree. Now if you ease up close, with luck you won't chase it off, but you will elicit instead its attempt to startle you away, by suddenly flashing its hind wing at you, which, depending on the species, will either be black and white, or black and red. Take a picture of it, and then the fun begins. In our old 1984 *Peterson Field Guide to Moths*, we would find pages of black underwings, and pages of red underwings. They were all pinned in the old way, with the wings pulled up, and we would have to practically stand on our heads trying to see the upper forewing as it would appear in our photo of the living moth, and try to guess which of the numerous very similar species it was. What a relief it was when the new *Peterson Field Guide to Moths* came out in 2012, showing pictures of living moths with their wings in normal position. That's when we found out they were still just as hard to identify. But that's all right: You can call them all "Underwings."

Luna Moth
(*Actias luna*)
Lepidoptera: Saturniidae

We can't imagine there is anyone who doesn't know the Luna Moth. This one has only recently emerged (eclosed is the technical term) from its cocoon down in the leaves. It is a male, as can be seen by his plumy antennae which have so many odor sensors on them he can identify a female Luna from a great distance with only a molecule of her scent. The main part of his life, really, was lived as a green, finger-sized caterpillar feeding on the leaves of a tree. Now, it has all come to a culmination. He has no eating mouthparts and will not feed again during the remaining few days of his life. He is essentially a mobile set of genitalia looking for a female to mate with and die, as she, after mating with him, is looking for a place to lay her eggs before she dies. There is, however, one nice little touch. They have to make these few days count, but they are big night-flying morsels of protein which a bat would love to pluck out of the air. This brings up their magnificent tails. They are not simply for beauty; It has only recently been learned that their swishings and swooshings around in flight actually scramble the bat's ability to echo-locate and thus make them harder to catch.[4]

---

[4] See http://www.sciencemag.org/news/2015/02/luna-moth-s-tails-fool-bat-sonar (accessed March 20, 2018).

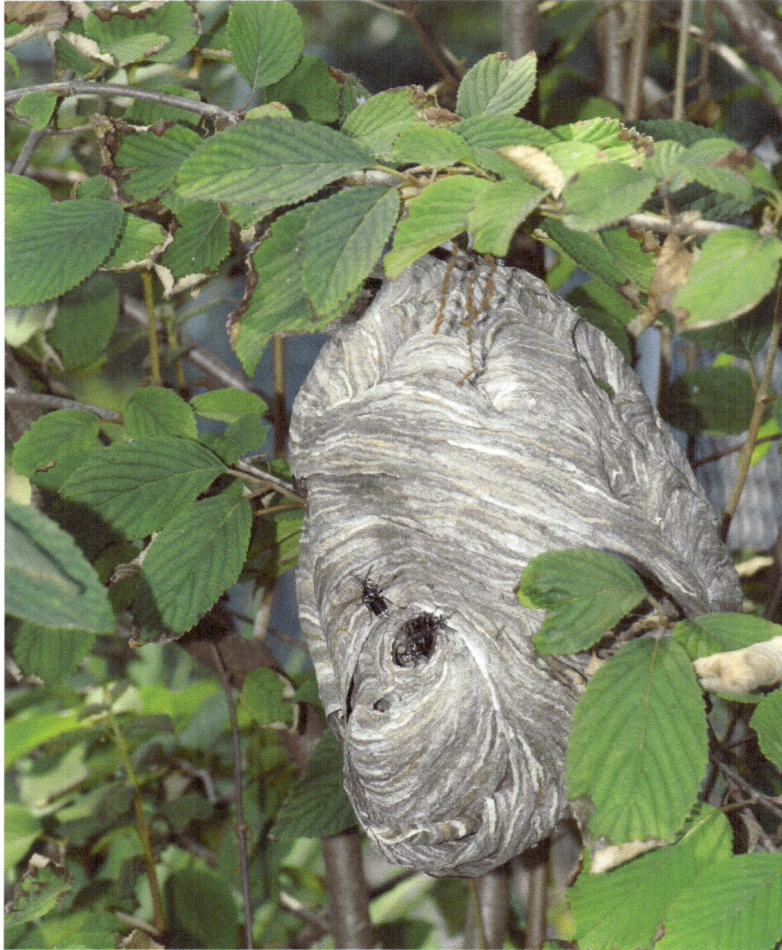

Bald-faced Hornet (*Dolichovespula maculata*)

Hymenoptera: Vespidae

Bald-faced Hornets make basketball-sized paper nests hidden in bushy foliage (coming into view when leaves are lost in autumn). We remember finding a nest in mid-winter (the inhabitants presumably dead and gone), bringing it home and setting it attractively in a corner of the study, only to learn, when the room had warmed up to spring-like temperature, that a couple of dozen pregnant queens waiting for next year were overwintering in the nest, and were now all emerging in very bad tempers.

These hornets are meat-eaters. The workers go out and catch horse flies, butterflies, or whatever insects they can, tear the prey into pieces, and bring it back to the nest. They are much faster than you might suspect. We saw one chase down a fleeing Least Skipper and catch it out of the air. And they have a special trick: On cold mornings they can, by shivering, warm themselves up to mammal-like temperatures. They then go out and look for insects that are too cold to fly away. You can see them flying along and smashing into every black spot they see on a leaf, in case it is a tasty insect.

Paper Wasp (*Polistes* spp.)

Hymenoptera: Vespidae

There are a number of species of Polistes paper wasps. They all have similar life behavior. They make the pizza-shaped open nests that hang from foliage, or often from under the eaves of your house. There is a large all-red species (*P. carolinus*) which we prevent from nesting around our house, since they tend to attack us, but the others (with various patterns of black, yellow, and red) are inoffensive and we enjoy watching them develop. The nest shown top right is in an early stage. The over-wintered pregnant queen started this nest by herself and raised the first few workers. This is the perilous time and many nests fail, but she has made it through. Up near the top of the picture you can see eggs inside the cells, and a newly hatched grub. Below, a fully developed grub is spinning a cocoon around itself. Then there are three closed cocoons developing. The dark smudge at the top of the nest is the queen herself, resting on the stem that supports the nest. She no longer has to risk her life going out all day foraging. Her job now is to stay there laying eggs while the others bring caterpillars and other insects to feed the babies like birds in a nest. There may be fifty or so workers by the end of the season.

Right is a mature *Polistes* nest in September, like the flight deck of an aircraft carrier, ready for combat. When we got under the nest, which is in the eaves of our house, they all snapped to attention, and if you look you can see that many of them are staring straight at us. This is a peaceful species, but we didn't go any closer.

At the end of the season the paper wasp queen and all the workers will die, but before they do they begin producing males and virgin queens. As the days get shorter, the virgin queens begin looking for sheltered places to hibernate—hollow logs or, particularly, insides of buildings, attics, or openings in the eaves. As they find these places, it turns out the males (identifiable by their white faces) are already waiting there to waylay them and won't let them in until they have mated. The males then die, and the pregnant queens winter in the shelters, emerging the following spring to start a new generation.

61

Thread-waisted Wasp (*Eremnophila aureonotata*)

Hymenoptera: Sphecidae

This is one of the thread-waisted wasps. These are solitary wasps. The general pattern for solitary (as opposed to social) wasps is, the adults feed on nectar from flowers, and they feed their larvae on insects or spiders that they have paralyzed to keep them fresh. Different species of wasp specialize in particular groups of prey. (For instance, some specialize in caterpillars, some in crickets, some in horse flies, and so on.) In the case of this wasp, it burrows a hole in the ground and fills the hole with a single large caterpillar, lays an egg on the caterpillar, covers the hole, and then goes off to find another caterpillar to put in another hole with another egg. It's rare that you witness that part of their life history, but if you look through a large bed of flowers covered with bees and wasps and flower flies nectaring, you will very commonly see these slender shiny black wasps with silvery-white marks on their shoulders. It will only be after you have looked very closely that you realize the male, like a permanent presence, is clinging to the female's back.

Eastern Carpenter Bee (*Xylocopa virginica*)

Hymenoptera: Apidae

Carpenter Bees are big bumble bee-like, shiny-black-abdomened bees with none of the good manners or public-spirit of bumble bees. Instead they are bullies, crashing into the flowers driving off all the butterflies and more interesting insects that you were just getting ready to photograph, so they can have the front row at the nectar. They are such bad citizens that instead of going into flowers the front way, thereby paying for the nectar by picking up pollen to deliver to the next flower they visit, they just as often go to the base of the flower on the outside and chew a hole to go directly to the nectar, bypassing the pollen. With those same powerful jaws, they make their nests by chewing tunnels in wood which, if it happens to be the shelves of your garage tool room, can do some damage. And finally, they are a bad influence on the real bumble bees that quickly learn to use the holes that have been cut at the back of the flowers, so the bumble bees themselves are saved getting totally coated with pollen. (An honest bumble bee citizen is pictured bottom right.) Carpenter Bees have their place in the system, but we're not always sure what it is.

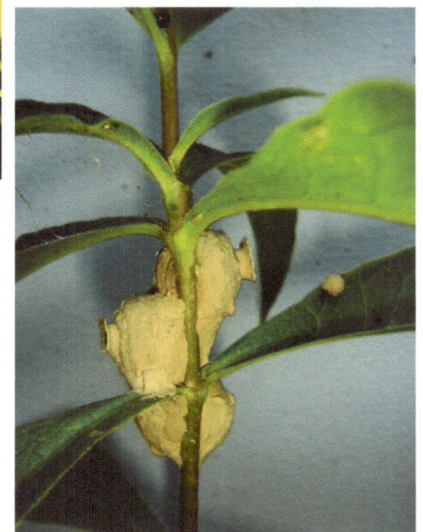

Potter Wasp (*Eumenes fraternus*)

Hymenoptera: Vespidae

Potter Wasps with their bulby bodies and natty black and yellowish-white markings specialize in catching caterpillars, which they paralyze and stuff into nests. When they have as many as they need, they lay an egg in the nest, and seal it up. Many of the solitary hunting wasps have a similar pattern. What makes this one special, and gives it its name, is the artistic quality of the nests they build. They are very attractive and delicately made clay pots.

Black-and-Yellow Mud Dauber (*Sceliphron caementarium*)

Hymenoptera: Sphecidae

This rather spectacular black and yellow thread-waisted wasp can often be seen at the edge of a mud puddle rolling up balls of mud, which it then carries off. These wasps are commonly called "dirt daubers" and we are sometimes asked if they really eat dirt. Of course, just like the swallows we see gathering mud at the same locations, they are gathering building materials for their nests. They make the globular mud nests we see under bridges, or on the walls inside open garages or shop buildings. They fill each long cell of the nest with spiders, laying an egg atop the spiders before sealing up the cell. The grub feeds on the spiders, makes a cocoon, then chews its way out and takes off to follow the family trade. This wasp is in the process of adding a dollop of mud to the cell it is preparing.

A nest accidentally got knocked down and broken, so we emptied it out and laid the contents on a piece of paper to see how many paralyzed spiders it held. Except for one jumping spider (Salticidae) it seemed to hold entirely crab spiders (Thomisidae, Philodromidae).

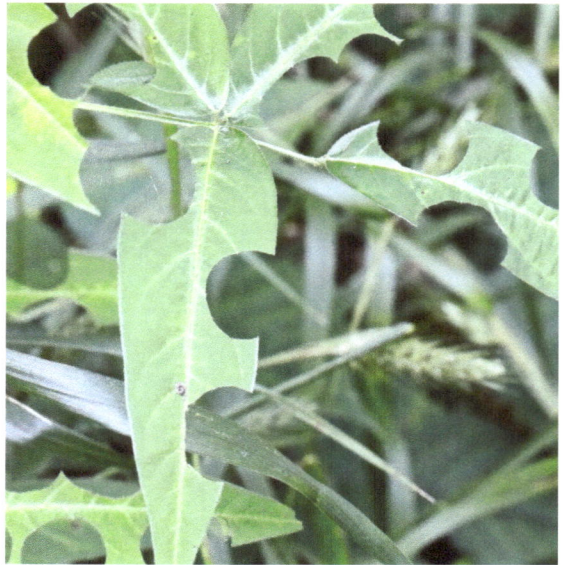

Leafcutter Bee (*Megachile* sp.)

Hymenoptera: Megachilidae

A very large buzz for a smallish bee will often give away a Leafcutter Bee foraging in nearby flowers. They are a solitary bee that makes a nest and stockpiles it with pollen, then lays an egg and seals the nest. They don't have pollen bags on their hind legs like social hive-living bees, but have their own ways of gathering and transporting pollen. They sort of squat down and run over the top of a flower, rubbing their bellies through the pollen, which sticks to hairs there. They are named for standing on leaves of the right consistency, and, holding a leg in one place like a compass, cutting a circular section of leaf (note their very powerful jaws), then carrying that section away to use as the walls of the cells they make for their young.

Above right is a Leafcutter Bee cutting a circle out of a leaf, and below, typical signs that Leafcutters have been at work.

Bottom left is one of the leaf-made nests we pulled out of a crevice after the bee was through using it. It is comprised of two cells, back to back, with exit holes at each end.

Sleeping Bee (*Melissodes* sp.)
Hymenoptera: Apidae

This is a large genus of solitary bees, including over a hundred species in this country. You can recognize them by their long antennae and yellow faces, and you can often spot a male holding a twig by his jaws as he sleeps. The females dig very complicated tunnels underground, prepare separate cells in the dirt, stock the cells with pollen and nectar, and lay an egg in each one. Every time they leave their nests, they fill the tunnels in with dirt. When they return, they dig them up again. Their whole life is work—tending to their tunnels and stocking them with food for their children.

The males, on the other hand, like those pictured here, are most famous for sleeping. They often gather with a number of other males on some small plant, hang from twigs by their jaws, and sleep the night away. No one is quite sure why they do this. During the day they chase after the females. This basically sums up their lives. But watch for them in your garden; it is not unusual to find one to a dozen sleeping out in plain sight, even during the day.

Ichneumon
(*Gnamptopelta obsidianator*)
Hymenoptera: Ichneumonidae

Ichneumons are a kind of parasitic wasp. There are thousands of kinds, extremely difficult even for experts to identify. But this particular one is easy: It is rather large, and unmistakable with its long bright orange antennae, and its equally long lank abdomen. It flies rapidly, close to the ground along the leaf litter, and when it sees a large caterpillar, it stabs an egg into the caterpillar, and flies on. The wasp grub grows inside the caterpillar, not eating vital tissues so that the caterpillar goes on living and feeding. The caterpillar finally forms a pupa, the wasp grub finishes eating the host and forms its own pupa inside, and it is a new Gnamptopelta that finally emerges.

Cicada Killer
(*Sphecius speciosus*)
Hymenoptera: Sphecidae

These big handsome wasps are solitary, which is to say, each female makes her own nest; but they are social in the sense that they like to nest close to each other. If that nesting place is your own lawn, you and your friends and neighbors might feel uneasy having a number of such formidable looking creatures flying back and forth in your yard. Luckily, they are inoffensive and unlikely to cause you any problems. The reason they are so big and powerful is that they need to be to catch cicadas, themselves big powerful insects, to feed their larvae. They sting the nerve centers to paralyze them, then take the cicadas back to the nesting sites to bury them in holes in the ground and lay eggs on them. Interestingly, if they bury one cicada with the egg, that will produce a male wasp. If they bury two cicadas with an egg, that will raise a female. We guess the male doesn't need as much nourishment, as he only has one job: to inseminate the female, whereas a female must make nests, catch cicadas, and raise several children.

The center picture is of a male, waiting atop a bush for a female to pass. He eagerly flies up every time something large and orange comes by. This particular male even flew up when he saw Norman's red nose, but quickly lost interest when he got close. To the left, a female catches a cicada.

Velvet Ant (*Dasymutilla* sp.)

Hymenoptera: Mutillidae

Velvet ants come in all sizes, but always with almost exactly this pattern of red and black. It is a well known warning sign: Insects with this pattern have a large stinger that delivers very painful stings. (In fact, one large species is known as the "Cow Killer.") A number of flies and beetles mimic the pattern, to get some of the respect these command from a potential predator. The velvet ant is actually a wingless wasp. It is wingless because it spends a lot of time burrowing underground, and wings are simply in the way if you are pushing through dirt. Many species of solitary wasps paralyze the insect they specialize in, and bury it at the bottom of a hole in the ground with their egg on top of it, which upon hatching will have the paralyzed insect to feed on. The velvet ant, saving itself a lot of trouble, finds one of these ready-stocked nests, digs down to it, and lays its own egg on top, so when it quickly hatches the new larva can eat the other wasp larva and its store of food. This female velvet ant pictured top right has her antennae facing downwards and quivering, sensing signs of activity below.

If the female velvet ant finds her victim most efficiently by walking along the ground, the male finds the female best by flying overhead, so he does have wings. Sometimes we are following a female, watching how she operates, when a male will fly down out of nowhere, grasp the female, and fly off with her. What could be more romantic than that!

(*Enoclerus* sp.)

Coleoptera: Cleridae

This beetle is one of dozens of species of insects that mimic velvet ants, taking advantage of the velvet ant's reputation for a having a very painful sting.

Braconid Wasps (*Cotesia* sp.)

Hymenoptera: Braconidae

If something white down in the brush catches your eye, bend down and look. You may see a stationary caterpillar with up to a hundred tiny white cocoons clustered on its back. It has been parasitized by a tiny (5 mm) braconid wasp which has used its ovipositor (a tube through which some insects lay their eggs) to inject dozens of eggs inside the caterpillar's body. The wasp larvae have devoured the caterpillar from inside, eating non-vital tissues at first, so that the caterpillar could continue eating and growing, but now for the purpose of feeding the wasp larvae, not the moth or butterfly the caterpillar might have become.

When these larvae were ready, they punched their way out and spun their cocoons. As you can see by the open caps, the cocoons have now hatched, and the adult wasps have departed. The caterpillar may still be alive for a while, but hasn't the energy left to make its own cocoon. It's a common fate for caterpillars, but actually the wasps are the heroes here, the great controllers of population. Without them (and the migrating warblers and all the other carnivores), caterpillars would overrun the world, eat everything green, and bring down their own environment.

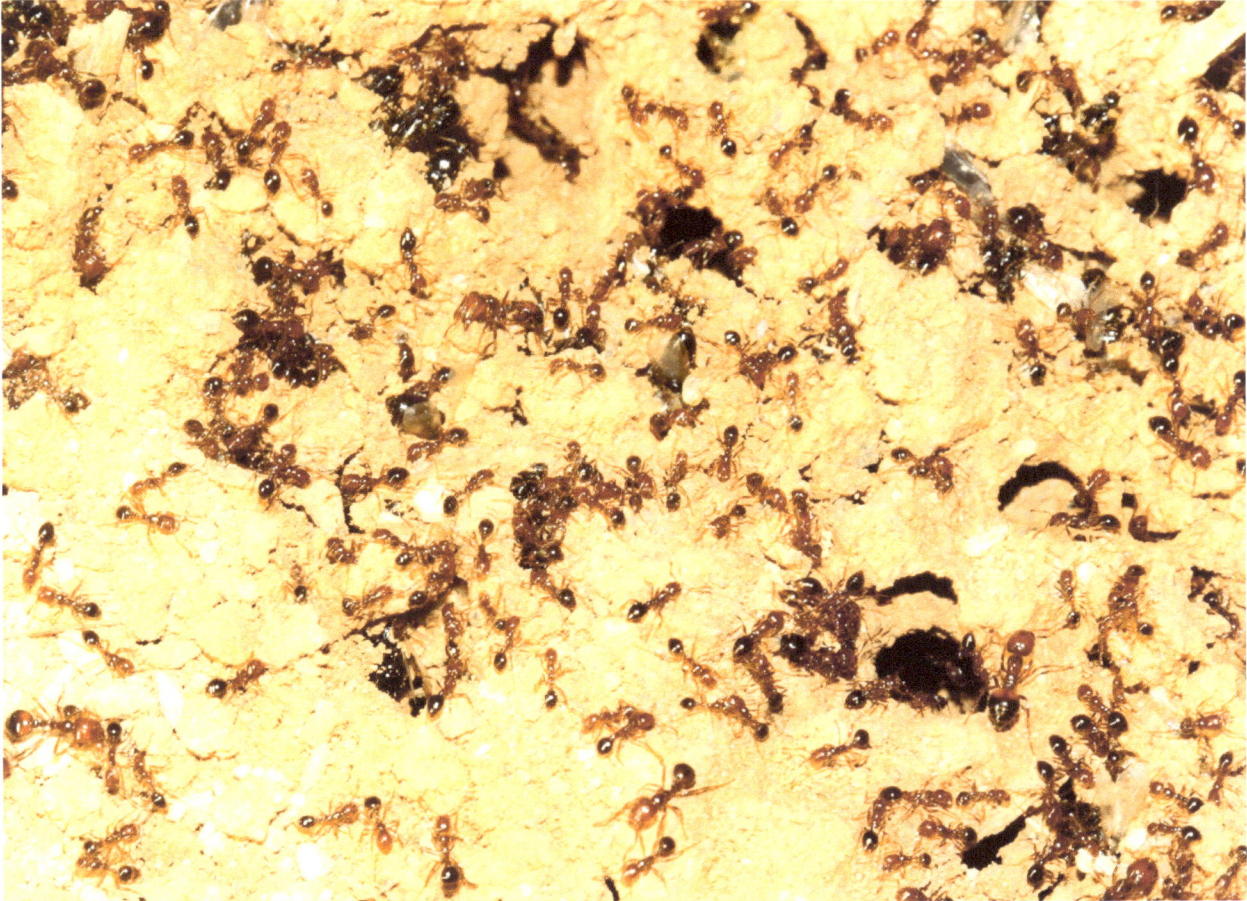

Red Imported Fire Ant (*Solenopsis invicta*)

Hymenoptera: Formicidae

These notorious ants, accidentally imported from South America, are a scourge throughout the southeastern United States. In places where there are dense populations you can drive along a road and see their mounds all across the fields. They wipe out other ant species, eat the babies of ground nesting birds, and give painful stings to the ankles of people who don't realize they are standing in a field swarming with them. They are spreading up from the south, but so far have only come a little over half way up Arkansas. They are accidentally introduced farther north in the potting soil of nursery plants, but it is thought and hoped they cannot endure the winters farther north than they already are. If you see a suspicious mound of dirt and want to know if it is a fire ant mound, there is an easy test (which, actually, is fun to do). Just give the mound a kick with your foot that shaves an inch or so off the top of the mound. Within two or three seconds it will be boiling over with ants, some carrying the young back underground, some beginning the job of re-covering the mound, and all prepared to attack if you don't quickly move a step away. What you are seeing is a good example of an (accidentally) introduced species which becomes a pest because it didn't bring with it the predators and parasites from its native home that held its numbers in check.

Crane Fly (Tipulidae sp.)

Diptera: Tipulidae

Crane flies are in the order Diptera, the "true" flies. The Latin word diptera means "two winged," and since the normal number for insects is four wings, this is a distinct difference. We might suppose this would mean they had diminished powers of flight, but the opposite is the case. Two wings are evidently more efficient, and the best flying Flies are superior to the best fliers in other orders. If you don't believe so, you should see robber flies snatching dragonflies out of mid-air (see later in this section).

Crane flies themselves are not among the great flyers of the fly world, but they are convenient examples to illustrate the traits of true flies. Note first that there are only two wings. They are the forewings. Where the hind wings would be is replaced by a little stalk with a knob on it. It is called a haltere, and is in fact a kind of gyroscope, stabilizing flies during the most complicated maneuvers.[5]

There are well over a thousand species of these "Daddy Longlegs" flies in the United States. They come in sizes as small as mosquitoes up to giants with a four-inch leg spread. Locally, the big ones here are called "skeeter hawks," and it is thought they hunt down and kill mosquitoes. The folk logic is, they look like mosquitoes, but they are much larger, just as, in folk medicine you need something that looks like the malady but is more powerful, to cure it. Crane flies actually are totally inoffensive and, with a few exceptions, vegetarian.

We often hear that there are more species of beetles than of practically all other living things. This truism is now being questioned. Beetles have always been popular with insect collectors and scientists alike, perhaps partially because with their hard bodies and often bright colors, they make such good specimens. Flies, with their often soft and fragile bodies, do not look as impressive in display cases. For whatever reason they are not collected as avidly, and so get less attention from entomologists, but in fact there may be upwards of a million species of flies that are waiting to be described. Flies, with their colors and shapes and complexities of behavior, may finally turn out to be the most numerous after all. They are our personal favorites.

Phantom Crane Fly (*Bittacomorpha* sp.)

Diptera: Ptychopteridae

If you are walking a path that follows a narrow often shadowed ditch that has thick emergent aquatic vegetation and two or three inches of water in low spots, that is a time to begin searching for Phantom Crane Flies (pictured right). You might be sharp-eyed enough to spot one hanging from a bit of vegetation, but most likely you will see a tiny movement in the shade that you won't believe at first (a phantom shape?). Then, if it crosses into a sunny spot, you will see the white bands on the legs flickering. The displaying males fly about two feet above the ground and just sort of hang in space with their legs splayed out for maximum effect of the leg-bands catching the dappled light, everything else remaining invisible. It worked for this male: the female has flown up to meet him.

[5] See https://www.sciencedaily.com/releases/2015/11/151125125145.htm (accessed March 20, 2018).

(*Tabanid* sp.)

(*Tabanus Quinquevittatus*)

Horse Flies (Tabanidae)

Diptera: Tabanidae

These are a couple of the numerous kinds of horse flies. You may think these remorseless blood-suckers have little to recommend them (except that they often have beautiful eyes), but in fact they are a major food for fish and flycatchers, and several species of wasps and other predatory insects. On the negative side, females have sharp knife-like mandibles that actually slice you open to make your blood flow. The males only have a soft tongue, and drink juice from plants.

The horse flies typically lay their eggs on foliage over water. When the eggs hatch, the aquatic larvae drop into the water, where they complete their development.

Fruit Fly (*Rhagoletis* sp.)

Diptera: Tephritidae

Here is a particularly clever scheme for a weaponless creature to pretend to be something dangerous. The Fruit Fly above is facing left, but the wing markings make it look astonishingly like a jumping spider facing right. In fact when being stalked by a jumping spider, this fly will sometimes walk backwards towards the spider, and make the motions of a spider looking for a fight, making the real spider back off.[6]

At right is a typical jumping spider for comparison.

---

[6] See Foelix, R.F. 2011. Biology of Spiders. Oxford University Press, New York, New York, 314.

Wood Midge (Cecidomyiidae sp.)

Diptera: Cecidomyiidae

Sometimes when you are driving country roads, you will see ghostly white smoke-clouds hanging in the air over the ditches. We drove by these for years, knowing it was swarming insects of some kind, but only recently decided to stop and see what they were. We stood watching the clouds while they constantly shifted position and configuration in the patterns made famous by "murmurations" of starlings and shorebirds. It was quite beautiful. But a surprise, when we approached more closely, was to discover how absolutely tiny the individuals were, less than 2 mm long. Look at this picture of one on the tip of a finger to get some sense of scale.

We searched through BugGuide[7] and finally learned that they were a species of Wood Midge. Their larvae feed on decaying wood. It is assumed that the swarms have something to do with mating, though this evidently has not been observed, and we noticed that this particular swarm seemed to consist entirely of males (identifiable by the claspers at the end of their abdomens). Many species of these tiny flies have yet to be described. There are so many insects, and so much still to be learned!

[7] This marvelous resource is discussed more fully in the bibliography.

Leaf Miners (Agromyzidae)

Diptera: Agromyzidae

Some groups of flies live the larval part of their lives between the top and bottom layers of a leaf, which protects them from predators. They eat the part of the leaf that is directly in front of them. As they progress, they make a tunnel just their width, and as they grow bigger, the tunnel grows wider. Here, you can see at a glance this individual's entire larval life from where it came out of the egg to the point where it abandoned the leaf to form a pupa. There are also moths and beetles that "mine" the leaves. Other kinds of leaf miners make blotches or other markings, but these are the most fun to observe.

Moth Fly (Psychodidae)

Diptera: Psychodidae

This tiny (4 mm) creature gets its name because it resembles a moth, but in fact is a kind of fly (the size has been blown up drastically in this photograph). It is usually to be seen as a little dot on a bathroom or kitchen wall but you can make out the characteristic shape even in such a small creature. Its larvae live in drains and feed on whatever goop is trapped there. They do no harm (though some of their relatives are sand flies, biting insects that, in the tropics, spread serious diseases).

Pomace Flies, "Fruit Flies" (*Drosophila melanogaster*)

Diptera: Drosophilidae

Drosophila melanogaster, commonly called "fruit flies" (though technically they are from another family of flies that also happen to be attracted to rotting fruit), are the famous laboratory animals liked by experimenters because they are cheap and easy to keep, go through myriad generations every year, and have chromosomes that are particularly easy to work with. We don't like them as much as the scientists do. The problem being, one of us (Norman) loves bananas, particularly super ripe ones that look more and more like black tar sitting on our shelves during the summer, which is just what they lay their eggs on. We soon find we are breeding a little cloud of them that follow him around the house when he is eating his peanut butter and banana sandwiches, and drown themselves regularly in the dregs of his wine glass. We were surprised that they are so fond of lemons. You evidently attract even more of these flies with sour than with sweet. Perhaps that is why they are also called vinegar flies

Freeloader Flies (Milichiidae)

Diptera: Milichiidae

A House Spider has caught a big Stink Bug, and as if they were waiting around for just such an event, a mob of tiny Milichiid or "freeloader" flies has moved in for its share of blood and dripping juices. The spider evidently has no idea they are there, and completely ignores them.

We have seen them riding on the backs of predatory insects, such as robber flies, to be on the spot when a kill is made.

Midge (Chironomidae)

Diptera: Chironomidae

Midges (unrelated to the wood midges shown earlier) have aquatic larvae and can emerge from the water as adults in billions, feeding fish and any other creatures around small enough to think them worth the effort. We can barely see them they are so small, and what we can see makes them look like small mosquitoes. But if we take a photograph of one with our macro camera, and blow it up (usually the best way to identify an unfamiliar insect), we will see that they rest with their front legs put up over their heads and extended in front of them, opposite from mosquitoes that

sometimes raise their hind legs in the air. Also these Chironomid midges are completely harmless. The female has ordinary antennae, but the male, as here, has wonderful full and fluffy antennae, sensitive enough for him to recognize his species of female in the great flying nighttime swarms.

The tiny underwater Midge larvae wave around like tropical eels, finding bits of detritus to feed on. They constitute an important fish food. (Note a few swimming clams in the foreground.)

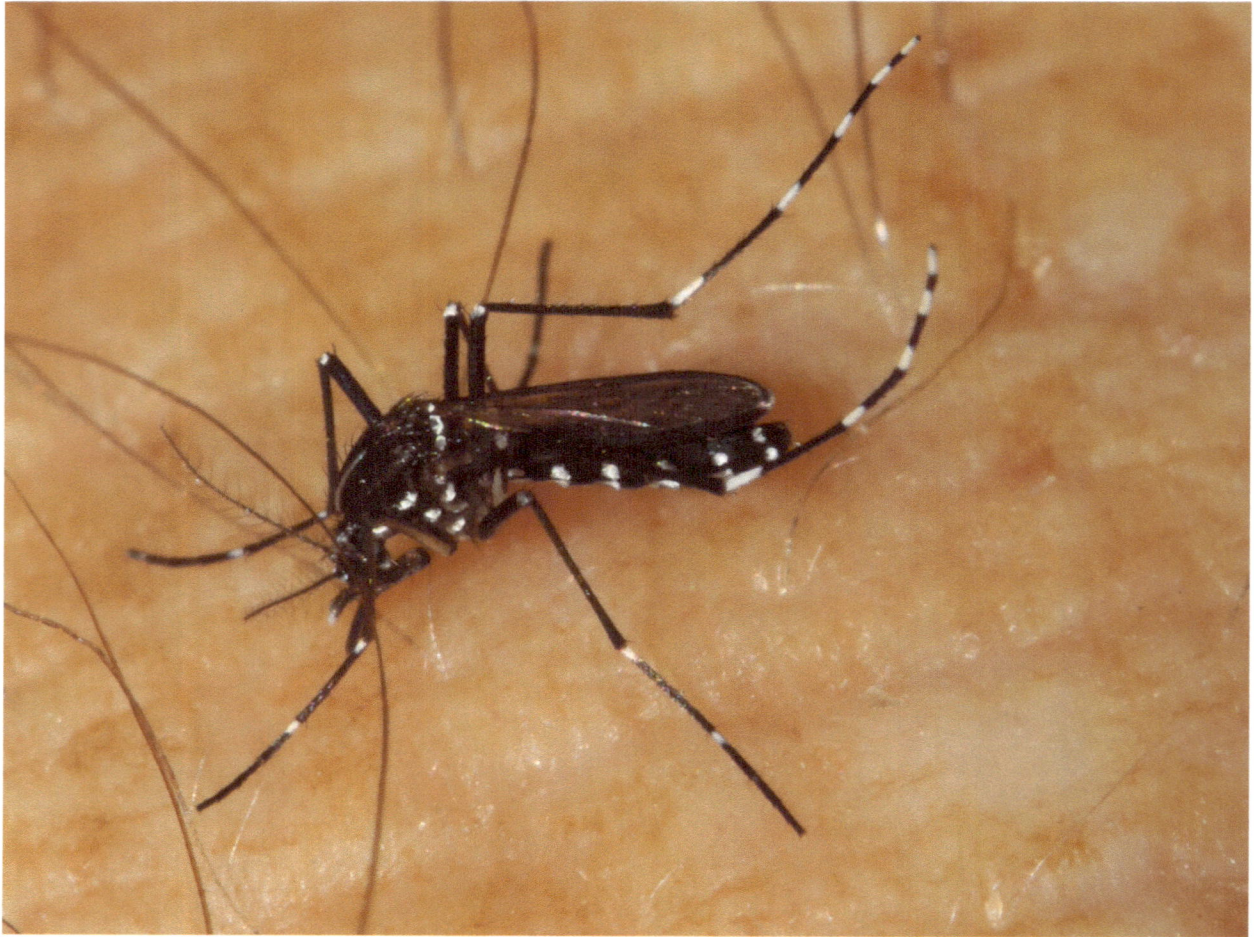

Asian Tiger Mosquito (*Aedes albopictus*)

Diptera: Culicidae

Aedes mosquitoes were much in the news recently, though we expect most people could not identify one if they saw it. In fact, they are so small most people are unlikely even to see them. They are very pretty mosquitoes with their black and white markings, but they are half the size of most mosquitoes, and one biting you looks like a tiny black speck on your skin. *Aedes albopictus*, the Asian Tiger Mosquito, was introduced to this country about thirty years ago in a shipment of used tires (they breed inside tiny bits of rain water collected in the tires, or for that matter in any tiny cup or broken bottle in trash that has liquid trapped inside it), and they are now established in most of the eastern United States. If you have a digital camera with macro capability, it is easy to photograph them and enlarge the photograph to see fine details. In the case of the individual pictured here, the fact that it has a narrow white stripe down the center of the thorax identifies the species, and separates it from *Aedes aegypti*, the notorious vector for the Zika virus, which is identical, except does not have the stripe. A *aegypti* is now established across the southern border of the country from Florida to California, with only occasional occurrences in Arkansas.

Elephant Mosquito (*Toxorhynchites* sp.) female

Diptera: Culicidae

If you see this extra big bright golden-colored mosquito, withhold your slapping reflex. Note the long, drooping proboscis, where you think there would be a stiff beak. This mosquito does not bite and suck your blood, but instead drinks innocently from plant nectar. That's one reason not to swat it. The second is that it is a very attractive insect. The third (and this is almost too good to believe), the larvae of this mosquito feed on the larvae of other, blood-sucking mosquitos. Perhaps that's why the female does not need a blood meal before laying her eggs: The larvae supply their own protein supplement. They are actually used to control mosquito populations. Maybe this big mosquito, half-way to being crane fly size, is the one that should be called a "skeeter hawk."

Long-legged Fly (Dolichopodidae)

Diptera: Dolichopodidae

Here, at left, is a is a very small fly that manages to call attention to itself first by generally sitting out in the open on top of a leaf (usually in a wetland area), and then by its handsome metallic colors, and third by the way that it races around on the leaf, apparently just for the pleasure of doing so. Just watching it will make you smile.

However, it is seriously at work, protecting its bit of territory from other males. And it is a fierce predator of midges, springtails, and other tiny creatures. Larvae of many species specialize on the larvae of bark beetles (*Dendrictinus* spp.), creatures which can do terrific damage to forests.[8]

Flower Fly (*Toxomerus* sp.)

Diptera: Syrphidae

If your garden has a long border of flowering plants the flowers will sometimes be covered with flying insects, bees, butterflies, wasps, and a large number of small yellowjacket-like flies. These latter are flower flies in the family Syrphidae. There is a large variety of them, and as adults, they are very important pollinators of flowers. There is another service they perform: The adults feed on nectar, but the larvae of many species feed voraciously on Aphids, and are major controllers.

---

[8] See http://www.nadsdiptera.org/Doid/Dolichar/Dolichar.htm (accessed March 20, 2018).

Flower Fly (*Milesia virginiensis*)

Diptera: Syrphidae

If this very pretty fly is anywhere around, you will see it, or, rather, hear its loud buzz. The larvae of this flower fly feed on decaying wood, and if the male finds an old log or the rotting center of a dead tree, he will take it over as his territory, and use that as a draw to get females to come to him, where they will have a nice place to lay their eggs. What makes the fly easy to see is that he constantly advertises that he has a tempting piece of wood to offer by hovering out in front of it. He at the same time shows off to the female his incredible endurance and flying ability by hovering in the air seemingly by the hour scarcely varying an inch from the spot where he began. Many species of flies in this large group, the Syrphidae, devote so much time to this behavior that they are collectively called "hover flies."

Syrphidae are harmless creatures, but like this one, many of them are patterned and colored like wasps and bees to fool predators into thinking they have venomous stingers.

Flower Fly (*Helophilus* sp.)

Diptera: Syrphidae

This is not one of the species that has predatory larvae. The larvae just feed on detritus in polluted water, but are still noteworthy. They are known as "rat-tailed maggots" because of their long snorkel tube which connects to the surface while they are feeding.

Below is one we set out on the edge of our birdbath for a moment so you could see it.

Flower Fly
(*Dioprosopa clavata*)

Diptera: Syrphidae

This Flower Fly has a different look from the others, but if you look at the very dark wing vein near the leading edge of the wing, that is one of the technical identification marks for the Syrphidae. The larvae of this one feed on Aphids.

Bee Fly (*Exoprosopa* sp.)

Diptera: Bombyliidae

Various species of Bee Flies will also be mixed in with the Flower Flies and others. These will be up in the flowers, but often will be seen flying along close to the ground. They always hold their long wings out almost at right angles to their body, and this is an easy way to recognize them. Their larvae are parasitic on ground-living insects of several families. With many species, the adult hovers over a hole in the ground and drops eggs down from the air.

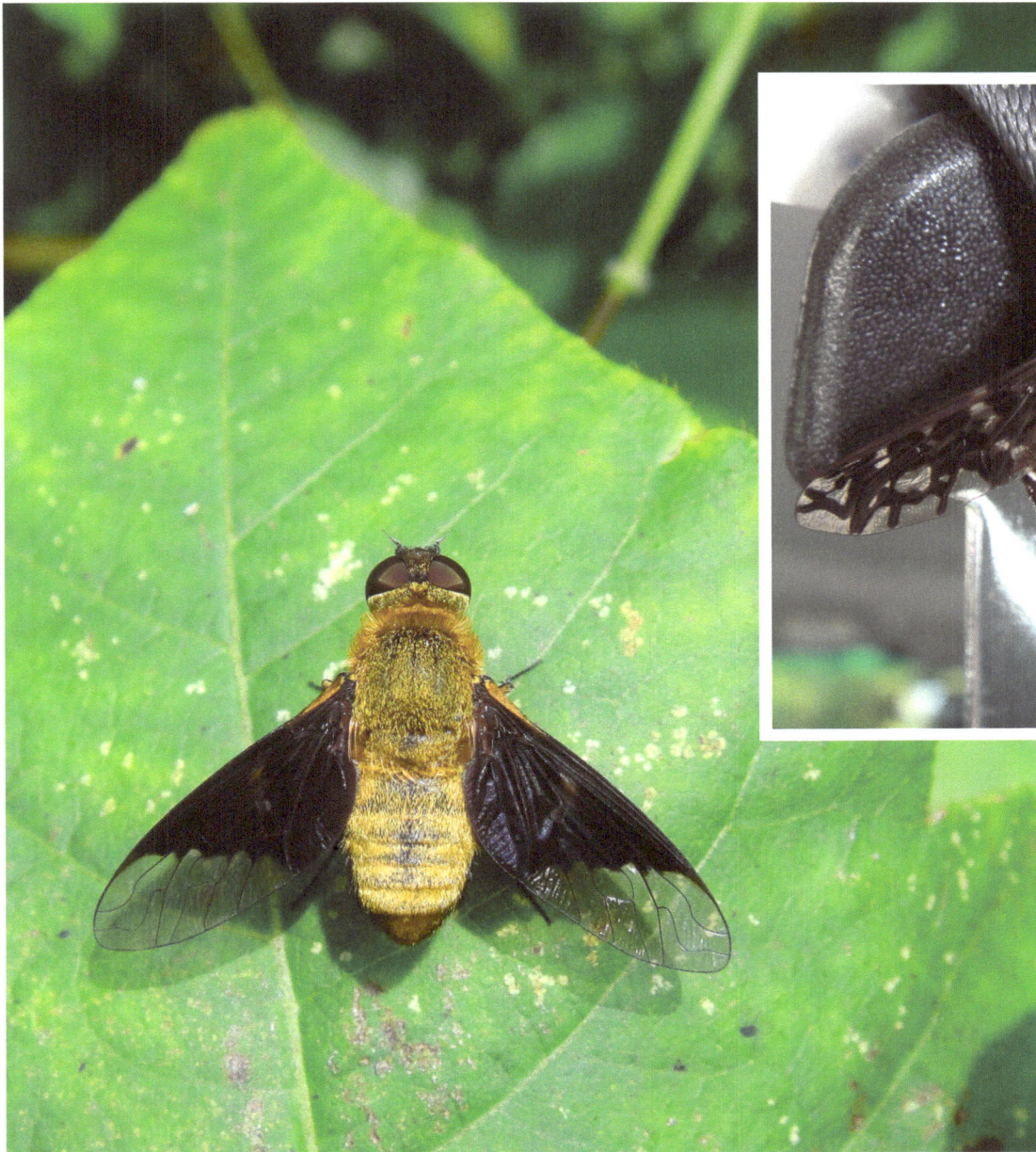

Bee Fly (*Chrysanthrax* sp.)

Diptera: Bombyliidae

The larvae of this stately fly are parasitic on scarab beetles.

Tiger Bee Fly (*Xenox tigrinus*)

Diptera: Bombyliidae

This large fly is distinctive enough that it has a common name. Its larvae are parasitic on larval Carpenter Bees.

Bee Fly (*Bombylius major*)

Diptera: Bombyliidae

Here is another very distinctive species. Note the long proboscis pointed down into the dirt. This species and many other bee flies fly with this long proboscis sticking straight before it, so that many observers mistake it for a very formidable beak. In fact, it is essentially a soft drinking straw for sucking nectar from plants.

Belvosia sp.

Gymnosoma sp.

Trichopoda sp. ("feather-legged fly")

Archytas sp.

Tachinid fly (Tachinidae spp.)

Diptera: Tachinidae

Often people think there is only one kind of fly, the house fly, a pest that crawls around on filthy things then comes over and lands on us, spreading diseases. There is such a thing as a house fly, and there are other flies that bite us, and so on. But the vast vast number of flies go on about their business and don't bother us at all, and in fact many do valuable work for the environment. There are, for instance, over a thousand kinds of Tachinid flies (Tachinidae) in North America, and they are major population controllers. You can often see the adults on flowers, where they also are important pollinators. They look like ordinary "flies" except most of them have very spiny abdomens. In fact, they are often called hedgehog flies for that reason. The adults sip nectar, but their larvae are internal parasites on a multitude of insects, mainly caterpillars. They have numerous methods for getting their eggs inside their hosts, and when the eggs hatch the tachinid larvae eat their hosts from the inside, carefully avoiding vital tissues to keep their hosts alive and feeding long enough for the tachinid larva to fully develop.

At left is a doomed caterpillar with a sprinkling of Tachinid eggs stuck to its skin.

Robber Fly (*Diogmites* sp.)

Diptera: Asilidae

Unless someone points them out to you, you are likely to overlook robber flies. There are over a hundred species in Arkansas in all sizes (from 3 mm to 50 mm) and shapes. They are powerful and charismatic predators, and often mimic wasps or bees, especially bumble bees. Once you get your eye in for them, you will see them sitting at the tip of a twig, or on a bare patch of ground, with an unobstructed view of the sky. Suddenly, they will fly up so rapidly they seem to vanish, only to reappear at the same spot an instant later, this time carrying some insect they have just snatched from the sky. They fly up like falcons, wrap their long spiky legs around their prey (often a wasp or other dangerous creature, even another robber fly), stab it with their beak loaded with neurotoxins and digestive enzymes, and return to earth to suck its juices. Those like the one pictured here, in the genus *Diogmites* (sometimes called "hanging thieves"), have the habit of getting under the foliage and hanging by their forelegs to eat their prey. You can recognize them by this curious behavior.

Robber flies have these features in common: widely separated eyes for good depth perception, a sharp beak, a sort of beard or mustache above the beak (which is thought to protect the eyes and mouthparts during encounters with dangerous prey), and long muscular spiky legs with hawk-like talons at the ends.

Robber Fly (*Laphria lata*)

Diptera: Asilidae

Many of the robber flies in the genus *Laphria*, such as this big bruiser, are very good mimics of bumble bees. They copy very exactly the hairiness and the aposematic coloration. One purpose for doing this is that it accords them protection from predators that don't want to be stung by a bumble bee. In that case, it is called "defensive mimicry." It must work, or there would not be so many insects in addition to robber flies that copy bumble bees. But there is another side to bumble bee behavior: While it is known that the stinger is there, it is also known that the bumble bee is so peaceable that it would never think to use its stinger unless it was under attack. Great numbers of flies, beetles, butterflies, and so on come to flowers and nectar calmly right beside bumble bees. A robber fly could take advantage of that, cozy up to an insect on the flowers beside it, and then grab it. That, technically, would be called "aggressive mimicry."

Robber Fly (*Laphria saffrana*)

Diptera: Asilidae

Here's another nice big Robber Fly. This one is in the same genus (*Laphria*) as a number of species that perfectly mimic bumble bees. But this particular one has decided to mimic a hornet, instead, and as if to underline how dangerous hornets are, it has what looks to us, anyway, like a death's-head skull on its thorax. The larvae of this one feed on beetle larvae that feed in decaying pine trees, so you find this species in pine forest. Often, in fact, the male will be sitting on a rotting pine log, waiting to waylay females that come to lay their eggs. We're sure this species is a fierce predator of local insects, but it seems to be rather friendly to people. We've often had one land on a boot, or on our clothing, and ride along with us when we are walking down woodland trails. This actually is something many robber flies do, and we imagine it's because people make good observation posts for a predator looking for insect prey (especially if that walker is also looking for insects!).

Robber Fly (*Promachus hinei*)

Diptera: Asilidae

This large striking Robber Fly, with its red legs, its red-brown thorax, and its light and dark brown banded abdomen, is a powerful predator of large insects found in woodlands. It is no slouch that can go up and catch a dragonfly in free flight and bring it down.

Robber Fly (*Cerotainia albipilosa*)

Diptera: Asilidae

Above is a tiny Robber catching a midge. Robbers don't all have to be enormous. This one looks like a dot sitting at the end of a leaf. There are lots of tiny prey animals and there need to be predators to catch them. This robber is only 5 or 6 mm long. There are other Robbers half that size.

Robber Fly (*Microstylum morosum*)

Diptera: Asilidae

But great big Robbers are necessary too. If you are in one of the remnant prairies in Arkansas, you won't want to miss this formidable predator. It's a robber fly nearly two inches long, which makes it, we imagine, not only the largest robber fly, but perhaps the largest fly around. Unlike most robbers, it does not sit in a certain spot and wait for suitable prey to fly over. This one actively hunts grasshoppers, streaking along close to the ground and frightening them up into flight, where it easily overtakes them and brings them down. The very first one we ever saw flew right up to us, circled Norman several times so fast he almost fell over trying to keep his eye on it, hung in the air in front of him and looked him over, decided Norman wasn't worth bothering about, and then went on about its business.

Aphids (Aphididae)

Hemiptera: Aphididae

Once upon a time, when our garden had an outbreak of aphids, we went into a sort of despair. At least we didn't buy all kinds of poisons, with labels saying "Harmless. Do not use around children, pets, or old people." We did spray our plants with soapy water, and all those other folk remedies. That was before we learned how interesting aphids could be. Now we are disappointed if we don't get an outbreak of them.

Here's what you need to do. Put a stool next to a plant covered with aphids, so you can sit comfortably. Use, if you have them, your close-focusing binoculars, or better, your camera with a Macro lens, and start observing. An Aphid is a little polyp of sugar water. It has no defense, and does not try to escape. There is a whole industry of insects set up just for the purpose of eating aphids.

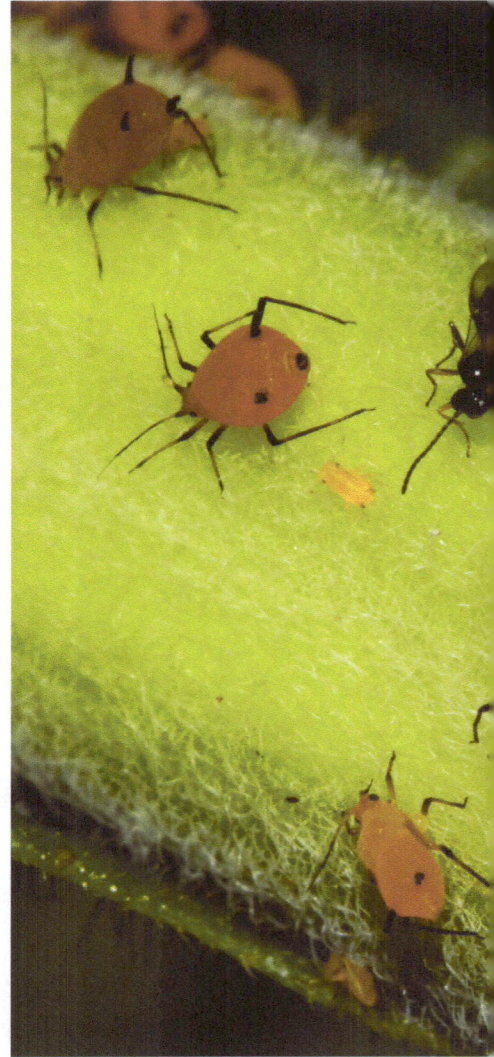

You might see a Braconid Wasp walking among the aphids, picking out a suitable victim. When it finds the one it wants, it will bend its ovipositor forward between its legs, charge the aphid, and inject an egg. Now that aphid will slowly turn brown as a larval wasp develops inside it, devouring the aphid, and finally pupating inside the shell of the aphid's skin. The next time you look, you may find all the aphids gone, leaving behind empty skins with neat circular holes in the top where the new wasps chewed their way out. If you had put down poison, the poison would have killed the wasps, and the Aphids would have re-invaded.

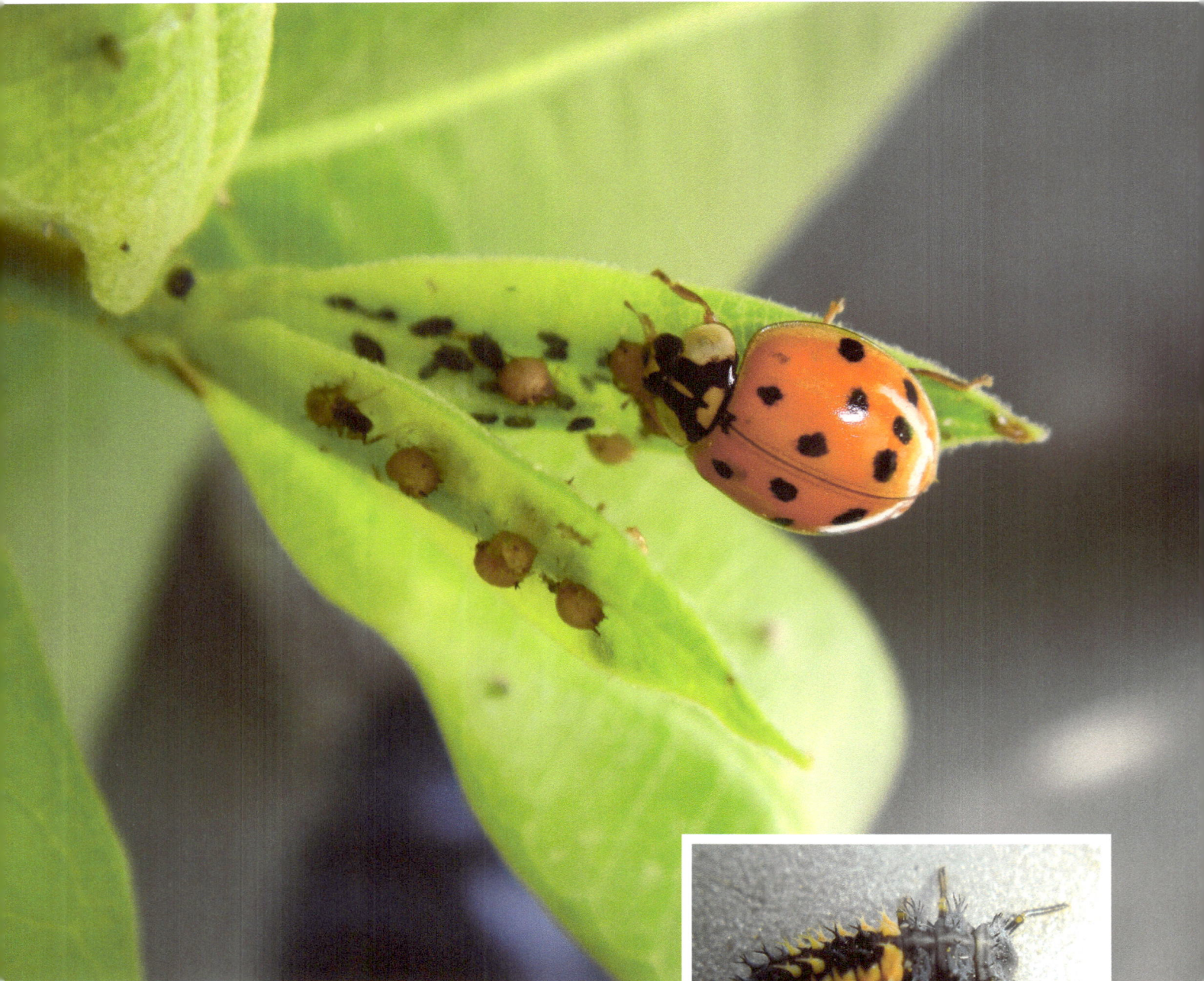

Also you will see lady beetles and lady beetle larvae
(which always look to us like miniature Gila Monsters)
eating the aphids.

Also this shaggy creature, with the wonderful name of Mealybug Destroyer (*Cryptolaemus montrouzieri*), is here in the act of devouring an aphid. The Destroyer is the larva of another species of lady beetle.

And also you might see the equally fierce larvae of Green Lacewings (which appear earlier in this book) and, here, of Brown Lacewings.

And, quite special, you might see the Harvester Butterfly (*Feniseca tarquinius*) laying its eggs next to a major infestation of aphids. Its slug-like caterpillars are the only purely carnivorous butterfly larvae in the country.

But most likely of all, you will see the larvae of the various Flower Flies, those pretty little bee and wasp imitators that do such a good job pollinating. Their larvae look, frankly, like decorated slugs, and they will be right in amongst the aphids, almost constantly gobbling them down. These are the major controllers.

Remember, however, if it was all destruction of aphids, with no defense at all, the aphids would be gone, and then all the specialist feeders on aphids would be gone. Just as, if all the predators were gone, the aphids would destroy all the plants, and then all the aphids would starve. Nature prefers the balance that keeps the largest number of forms of life, while still keeping a lid on those that would get out of control. So even aphids have their own non-violent peaceful defense. Here it is: With no need to mate with a male, aphids can pump out fully formed and ready-to-begin-eating babies all day long. In fact, believe it or not, that baby just emerging from its mother already has a baby forming inside it.

References and Further Reading

If this book has whetted your interest for insects, and you want to learn more, there are wonderful resources available. First, we would recommend that you get close-focusing binoculars (binoculars that focus at least as close as four feet, the closer the better). Bird watchers use binoculars that focus at around twenty feet away giving them clear enough images of birds that they they can make out the subtle details of pattern a bird field guide will tell them to look for to distinguish between species. Insects are of course much smaller than birds, but insects will usually allow you to get much closer to them. With your close-focusing binoculars (which, by the way, still focus on infinity, so they can be used for birdwatching as well) you can get images equivalent to the bird watcher's images of birds, and the new insect field guides are keyed to the details you can see through the binoculars.

For the next step, for a few hundred dollars you can get one of the small digital cameras that focus as close as an inch away (and be sure you get one that close; some cameras say close-up when they mean three feet). If you take a close-up picture of an insect, you can blow it up on your computer screen and compare it with the pictures in the field guides. That makes identification easier, but is also an enjoyable hobby in itself, catching images of these sometimes weird and often beautiful creatures.

You can go online to **BugGuide.net**, a site hosted by the Iowa State University Department of Entomology, which over the years has accumulated more than a million images of insects and spiders which curious people like you have submitted, and which experts in the various fields have volunteered to identify. You can browse their pages looking for the insect you have seen, and if you don't find it, you can submit your picture for their help. (http://bugguide.net/node/view/15740).

There are fine field guides to various popular groups of insects the way they appear alive in the field, and new ones are coming out constantly, as this is such a growing field. First of all you might get:

Eaton, Eric R., and Kenn Kaufman. 2007. *Kaufman Field Guide to Insects of North America*. Houghton Mifflin, New York, New York.

This has the largest general selection of insects in the easiest to use format of any pictorial field guide to North American insects. You can do very well with just this one field guide.

If you want to go into greater depth, there are excellent guides to single groups of insects. Here are some we use:

Beadle, David and Seabrooke Leckie. 2012. *Peterson Field Guide to Moths of Northeastern North America*. Houghton Mifflin Harcourt, New York, New York.

Capinera, John L., Ralph D. Scott, and Thomas J. Walker. 2004. *Field Guide to Grasshoppers, Katydids, and Crickets of the United States*. Cornell University Press, Ithaca, New York.

Evans, Arthur V. 2014. *Beetles of Eastern North America*. Princeton Field Guides, Princeton University Press, Princeton, New Jersey.

Fisher, Brian L. and Stefan P. Cover. 2007. *Ants of North America: A Guide to the Genera.* University of California Press, Berkeley, California.

Glassberg, Jeffrey. 2017. *A Swift Guide to Butterflies of North America* (Second Edition). Sunstreak Books, Morristown, New Jersey.

Lavers, Norman. 2007. The Robber Flies of Crowley's Ridge, Arkansas. https://normanlavers.net/ So far as we know, Norman's is the only field guide to robber flies anywhere.

Paulson, Dennis. 2011. *Dragonflies and Damselflies of the East.* Princeton Field Guides, Princeton University Press, Princeton, New Jersey.

Pearson, David L., C. Barry Knisley, and Charles J. Kazilek. 2006. *A Field Guide to the Tiger Beetles of the United States and Canada.* Oxford University Press, New York, New York.

Raney, Herschel. Random Natural Acts. http://www.hr-rna.com/RNA/index.htm. Lots of photographs and links to useful information about invertebrates mainly in Arkansas.

Spencer, Lori A. 2014. *Arkansas Butterflies and Moths* (Second Edition). Ozark Society Foundation, Little Rock, Arkansas.

Wagner, David L. 2005. *Caterpillars of Eastern North America.* Princeton Field Guides, Princeton University Press, Princeton, New Jersey.

Here are some informative and entertainingly written books about insects in general:

Berenbaum, May R. 1995. *Bugs in the System: Insects and Their Impact on Human Affairs.* Helix Books, Reading, Massachusetts.

Evans, Howard Ensign. 1973. *Wasp Farm.* Anchor Press, Doubleday & Company, Inc., Garden City, New York.

Goulson, Dave. 2013. *A Sting in the Tail: My Adventures with Bumblebees.* Picador, New York, New York.

McAlister, Erica. 2017. *The Secret Life of Flies.* Firefly Books, Buffalo, New York.

Waldbauer, Gilbert. 2004. *What Good are Bugs? Insects in the Web of Life.* Harvard University Press, Cambridge, Massachusetts.

Zuk, Marlene. 2011. *Sex on Six Legs: Lessons on Life, Love, and Language from the Insect World.* Houghton Mifflin Harcourt, New York, New York

About the Authors

Cheryl Lavers grew up in Porthcawl, on the south coast of Wales, in a family of nature lovers. Her earliest memories are of walks in the countryside to see birds and wildflowers, especially the delicate terrestrial orchids of the nearby sand dunes. She studied painting at Bath Academy of Art and since coming to the U.S. has painted birds and plants, most recently concentrating on the wonderful variety of leaves found in Arkansas. Cheryl has enjoyed exhibiting her bird and botanical paintings.

Norman Lavers taught English all over the country before landing at Arkansas State University in Jonesboro, where he taught for twenty-five years. He grew up in Berkeley, California, and from the time he was a child brought home every creeping and crawling thing he saw and kept it as a pet.

Norman interrupted his study of English Literature to spend a year traveling in Europe and North Africa. On that trip, he met a young art student and happened to mention that he had heard the European robin was quite different from the American robin, and that he especially wanted to see one. She said, "Wait a minute. There's one right here." He proposed to her a week later. The Lavers have just celebrated their 50th wedding anniversary.

When Norman retired from teaching in 2000, the Lavers began studying and photographing the fantastic insects of Arkansas full time, sharing their enthusiasm for insects through presentations to Master Naturalists and Arkansas Audubon Society Adult Ecology Workshops, as well as other groups. This book is the result of the nearly twenty years the couple has spent observing, enjoying, and teaching others about insects in the wilds and gardens of Arkansas.

www.ingramcontent.com/pod-product-compliance
Lightning Source LLC
Chambersburg PA
CBHW060811270326
41928CB00003B/55